# Advances on Tensor Analysis and their Applications

*Edited by Francisco Bulnes*

Published in London, United Kingdom

IntechOpen

*Supporting open minds since 2005*

Advances on Tensor Analysis and their Applications
http://dx.doi.org/10.5772/intechopen.75333
Edited by Francisco Bulnes

Contributors
Pınar Kirezli Uludağ, Daniel Condurache, Rodrigo Eustaquio, Isaias Martinez, Dr. Francisco Bulnes, Omar Zamudio, Edgar Navarro, Dmitry Vladimirovich Nikushchenko, Valery Pavlovsky, Francisco Bulnes, Augusto Melgarejo, Claudia Ruscitti, Laura Langoni



First published in London, United Kingdom, 2020 by IntechOpen
IntechOpen is the global imprint of INTECHOPEN LIMITED, registered in England and Wales, registration number: 11086078, 7th floor, 10 Lower Thames Street, London, EC3R 6AF, United Kingdom
Printed in Croatia

British Library Cataloguing-in-Publication Data
A catalogue record for this book is available from the British Library

Additional hard and PDF copies can be obtained from orders@intechopen.com

Advances on Tensor Analysis and their Applications
Edited by Francisco Bulnes
p. cm.
Print ISBN 978-1-83962-555-8
Online ISBN 978-1-83962-556-5
eBook (PDF) ISBN 978-1-83962-557-2

For EU product safety concerns:
IN TECH d.o.o., Prolaz Marije Krucifikse Kozulić 3, 51000 Rijeka, Croatia, info@intechopen.com or visit our website at intechopen.com.

# Meet the editor

Dr. Francisco Bulnes has a PhD in Mathematical Sciences from Instituto de Matematicas (IM/UNAM). He also has two post-doctorate degrees in Mathematics from Cuba and Russia. He is director of the International Advanced Research in Mathematics and Engineering Centre in Mexico (IINAMEI). He is also editor-in-chief and reviewer of several mathematics and physics journals as well as a member of various international science committees. Since 2009, Dr. Bulnes has been head of the Research Department, TESCHA. He has published more than 100 papers in mathematics and physics journals, and authored several books. Dr. Bulnes is recognized in Eastern Europe, Asia, and Arab countries for his many theories, theorems, and math objects. He has received numerous recognitions and badges from universities and governmental and nongovernmental organizations. His biography has been included in numerous archives of great mathematicians. He is a distinguished member (JCFM) of the Czech Republic Mathematics Society.

# Contents

# Preface

Generalization of the concepts of scalar, vector and matrix, which are independent of any elected coordinates systems, brought forth the concept of tensor. A tensor is a mathematical entity born of the invariance idea of the mentioned concepts in any election of coordinate systems and to any coordinate system transformation.

The concept has been of great importance in describing the invariance of physics laws with respect to any coordinate inertial reference framework where physics phenomena are measured. Likewise, gravitation theory is an example of the importance of describing physics laws through their invariants. Theories like general and special relativity brought forth diverse Einstein summation laws and other properties from Levi-Civita calculus, contributing to special tensors inside the Riemannian structure, which best describes the phenomena of the universe and their relations. Such is the case of Riemann tensors, Pseudo-tensors, and the different curvature tensor types arising from theories on torsion field, Cartan-Einstein theories, or supersymmetries in quantum mechanics. A more mathematical focus on tensors considers the multilinear forms and the tensor product of the vector spaces, which have more relevance to tensor applications of Hilbert spaces, for example, in QED and quantum mechanics.

Applications in classical mechanics, electrodynamics, quantum mechanics and communication theory are well developed through tensors. In quantum communication theory and parallel geometries to Riemannian geometry, such as twistor geometry, spinors and twistors are considered new interpretations of the tensors in fields and waves.

**Dr. Francisco Bulnes**
Professor,
IINAMEI, Director

Research Department in Mathematics and Engineering,
TESCHA,
Mexico

# Section 1

# Fundamentals

Chapter 1

# Bilinear Applications and Tensors

*Rodrigo Garcia Eustaquio*


## Abstract

In this chapter, a theoretical approach to the vector space of tensor of order 3 and the vector space of bilinear applications will be presented in order to present an isomorphism between these spaces and several properties about tensor and bilinear applications. With this well-defined isomorphism, we will present how to calculate the product between tensor of second derivatives and a vector, where such a product is used in several numerical methods such as Chebyshev-Halley class and others mentioned in the introduction. In addition, concepts on differentiability are presented, allowing a better understanding for the reader about second-order derivatives seen as a tensor.


**Keywords:** tensor, bilinear application, isomorphism, second derivative, inexact tensor-free Chebyshev-Halley class

## 1. Introduction

Frequently, discretization of mathematical models demands solving a system of equations, which is generally nonlinear. Such mathematical problems might be written as

$$\text{find } x^* \in \mathbb{R}^n \text{ such that } F(x^*) = 0 \tag{1}$$

where $F : -\mathbb{R}^n \to \mathbb{R}^n$.

There exist iterative methods for solving (1) that have cubic convergence rate, for instance, the methods belonging to the following class of methods named Chebyshev-Halley class, which was introduced by Hernández and Gutiérrez in [1]:

$$x^{k+1} = x^k - \left[I + \frac{1}{2}\mathcal{L}(x^k)\left(I - \alpha\mathcal{L}(x^k)\right)^{-1}\right]J_F(x^k)^{-1}F(x^k), \tag{2}$$

for all $k \in \mathbb{N}$, where

$$\mathcal{L}(x) = J_F(x)^{-1}\mathcal{T}_F(x)\left(J_F(x)^{-1}F(x)\right), \tag{3}$$

and $J_F(x)$ and $\mathcal{T}_F(x)$ denote the first and second derivatives of $F$ evaluated at $x$, respectively. The parameter $\alpha$ is a real number and $I$ is the identity matrix in $\mathbb{R}^{n \times n}$.

Discretized versions of Chebyshev-Halley class have already been considered in [2] in such a way that the tensor of second derivatives of the function $F$ was approximated by bilinear operators. A tensor is a multi-way array or multidimensional matrix. A generalization of the Chebyshev-Halley class (2) where no second-order derivative information is required but that also has cubic convergence rate, named inexact tensor-free Chebyshev-Halley class, was introduced by Eustaquio, Ribeiro, and Dumett [3]. Other families of iterative methods with cubic convergence rate were extensively described in Traub's book [4].

Several alternatives exist for the product of the tensor of second derivatives of $F$ by vectors [5–8], and this needs to be elucidated.

The aim of this chapter is to present concepts and relationships between tensors of order 3 and bilinear applications, in order to relate them to the second derivative of a two-differentiable application. We will see later that given the vectors $u, v \in \mathbb{R}^n$, the $i$-th row of the matrix $\mathcal{T}_F(x)v$ is defined by $v^T \nabla^2 f_i(x)$, where $\nabla^2 f_i(x)$ is the Hessian of the $i$-th component of $F$ evaluated at $x$. The $i$-th component of the vector $\mathcal{T}_F(x)vu$ is defined by $v^T \nabla^2 f_i(x)u$.

## 2. Tensors

Tensors naturally arise in some applications, such as chemometry [9], signal processing [10], and others. According to [8], for many applications involving high-order tensors, the known results of matrix algebra seemed to be insufficient in the twentieth century. There were some workshops and congresses on the study of tensors, such as:

- Workshop on Tensor Decomposition at the American Institute of Mathematics which took place at the Palo Alto, California, 2004, organized by Golub, Kolda, Nagy, and Van Loan. Details in [11];

- Workshop on Tensor Decompositions and Applications, 2005, organized by Comon and De Lathauwer. Details in [12]; and

- Minisymposium on Numerical Multilinear Algebra: A New Beginning, 2007, organized by Golub, Comon, De Lathauwer, and Lim and which took place at the Zurich.

Readers interested in multilinear singular value decomposition, eigenvalues, and eigenvectors may consult as references [5–8, 13, 14]. In this text, we will focus our attention on tensors of order 3.

Let $I_1, I_2$, and $I_3$ be three positive integers. A tensor $\mathcal{T}$ of order 3 is an three-way array where its elements $t_{i_1 i_2}^{i_3}$ are indexed by $i_1 = 1, \dots, I_1$, $i_2 = 1, \dots, I_2$, and $i_3 = 1, \dots, I_3$ and the $n$-th dimension of the tensor is denoted by $I_n$, for $n = 1, 2, 3$. For example, the first, second, and third dimensions of a tensor $\mathcal{T} \in \mathbb{R}^{2\times4\times3}$ are 2, 4, 3, respectively.

Obviously, tensors are generalizations of matrices. A matrix can be viewed as a tensor of order 2, while a vector can be viewed as a tensor of order 1.

From an algebraic point of view, a tensor $\mathcal{T}$ of order 3 is an element of the vector space $\mathbb{R}^{I_1\times I_2\times I_3}$, whereas from the geometric point of view, a tensor $\mathcal{T}$ of order 3 can be seen as a parallelepiped [15], with $I_1$ rows, $I_2$ columns, and $I_3$ tubes. **Figure 1** illustrates a tensor $\mathcal{T} \in \mathbb{R}^{2\times4\times3}$.

In linear algebra, it is common to see a matrix through its columns. If $A \in \mathbb{R}^{m\times n}$, then $A$ can be viewed as $A = [a_1 \dots a_n]$, where $a_j \in \mathbb{R}^m$ denotes the $j$-th column of the matrix $A$. In the case of tensor of order 3, we can see them through fibers and slices. Hence follow the definitions.

**Definition 1.1.** A tensor fiber of a tensor of order 3 is a one-dimensional fragment obtained by fixing only two indices.

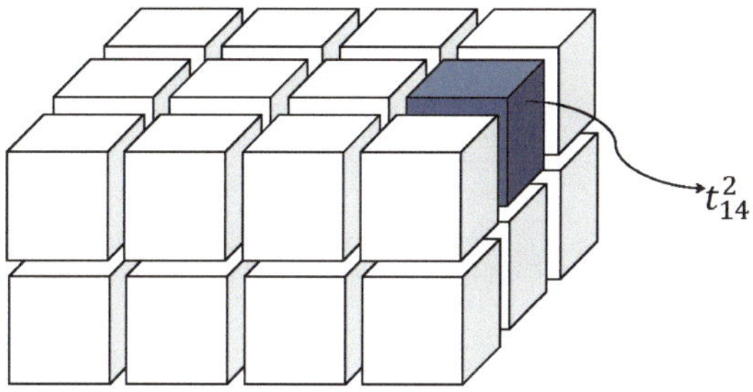


**Figure 1.**
*A tensor* $\mathcal{T} \in \mathbb{R}^{2\times4\times3}$.

**Definition 1.2.** A tensor slice of a tensor of order 3 is a two-dimensional section (fragment), obtained by fixing only one index.

Generally in tensors of order 3, a fiber is a vector and a slice is a matrix. We have three types of fibers:

- column fibers (or mode-1 fiber), where the indices $i_2$ and $i_3$ are fixed;
- row fibers (or mode-2 fiber), where the indices $i_1$ and $i_3$ are fixed; and
- tube fibers (or mode-3 fiber), where the indices $i_1$ and $i_2$ are fixed.

We also have three types of slices:

- horizontal slice, where the index $i_1$ is fixed;
- lateral slice, where the index $i_2$ is fixed; and
- frontal slice, where the index $i_3$ is fixed.

For example, consider a tensor $\mathcal{T} \in \mathbb{R}^{2\times4\times3}$ with $i = 1, 2$, $j = 1, 2, 3, 4$, and $k = 1, 2, 3$. The $i$-th horizontal slice, denoted by $\mathcal{T}^{i::}$, is the matrix

$$\mathcal{T}^{i::} = \begin{pmatrix} t_{i1}^1 & t_{i1}^2 & t_{i1}^3 \\ t_{i2}^1 & t_{i2}^2 & t_{i2}^3 \\ t_{i3}^1 & t_{i3}^2 & t_{i3}^3 \\ t_{i4}^1 & t_{i4}^2 & t_{i4}^3 \end{pmatrix},$$

the $j$-th lateral slice, denoted by $\mathcal{T}^{:j:}$, is the matrix

$$\mathcal{T}^{:j:} = \begin{pmatrix} t_{1j}^1 & t_{1j}^2 & t_{1j}^3 \\ t_{2j}^1 & t_{2j}^2 & t_{2j}^3 \end{pmatrix}$$

and the $k$-th frontal slice, denoted by $\mathcal{T}^{::k}$, is the matrix

$$\mathcal{T}^{::k} = \begin{pmatrix} t_{11}^k & t_{12}^k & t_{13}^k & t_{14}^k \\ t_{21}^k & t_{22}^k & t_{23}^k & t_{24}^k \end{pmatrix}. \tag{4}$$

**Figures 2** and **3** illustrate the three types of fibers and slices, respectively, of a tensor $\mathcal{T} \in \mathbb{R}^{2\times4\times3}$.

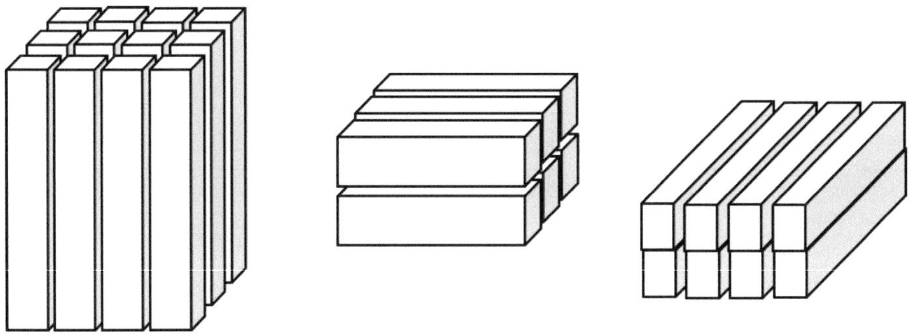

**Figure 2.**
*Columns, rows, and tube fibers, respectively.*

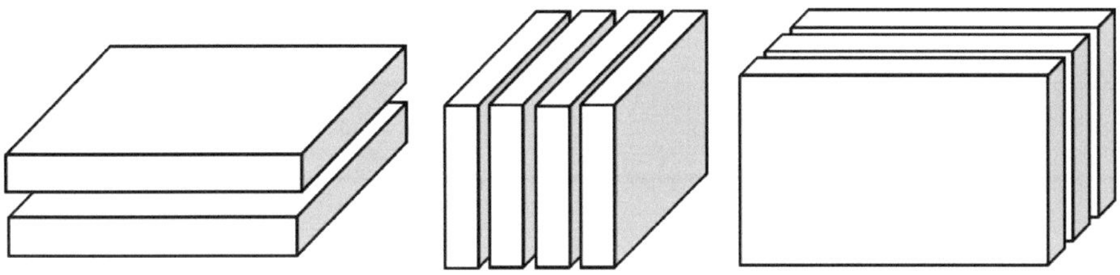

**Figure 3.**
*Horizontal, lateral, and frontal slices, respectively.*

## 2.1 Tensor operations

The first issue to consider in this subsection is how to calculate the product between tensors and matrices. It is well known from elementary algebra that given matrices $A \in \mathbb{R}^{m \times n}$ and $B \in \mathbb{R}^{R \times m}$, it is possible to calculate the product $BA$, because the first dimension (number of rows) of matrix $A$ agrees with the second dimension (number of columns) of matrix $B$, and each product element is the result of the inner product between rows of matrix $B$ and columns of matrix $A$.

The product between tensors of order 3 and matrices or vectors is a bit more complicated. In order to obtain an element of the product between a tensor and a matrix, it is necessary to specify what dimension of the tensor will be chosen to agree with the number of columns of the matrix, and each resulting element will be a result of the inner product between the mode-$n$ fibers (column, row, or tube) and the columns of the matrix. We will use the solution adopted by [8], which defines the product mode-$n$ between tensors and matrices and the solution adopted by [5] that defines the contracted product mode-$n$ between tensors and vectors.

The mode-$n$ product is useful when one wants to decompose into singular values a high-order tensor in order to avoid the use of the generalized transpose concept. We refer to [5, 7, 8, 13] for details.

**Definition 1.3.** (mode-$n$ tensor matrix product) The mode-1 product between a tensor $\mathcal{T} \in \mathbb{R}^{m \times n \times p}$ and a matrix $A \in \mathbb{R}^{R \times m}$ is a tensor

$$\mathcal{Y} = \mathcal{T} \times_1 A \in \mathbb{R}^{R \times n \times p}$$

where its elements are defined by

$$y_{rj}^{k} = \sum_{i=1}^{m} t_{ij}^{k} a_{ri} \quad \text{where} \quad r = 1, \dots, R, j = 1, \dots, n, \text{ and } k = 1, \dots, p.$$

The mode-2 product between a tensor $\mathcal{T} \in \mathbb{R}^{m \times n \times p}$ and a matrix $A \in \mathbb{R}^{R \times n}$ is a tensor

$$\mathcal{Y} = \mathcal{T} \times_2 A \in \mathbb{R}^{m \times R \times p}$$

where its elements are defined by

$$y_{ir}^k = \sum_{j=1}^{n} t_{ij}^k a_{rj} \quad \text{where} \quad i = 1, \dots, m, r = 1, \dots, R \text{ and } k = 1, \dots, p.$$

The mode-3 product between a tensor $\mathcal{T} \in \mathbb{R}^{m \times n \times p}$ and a matrix $A \in \mathbb{R}^{R \times p}$ is a tensor

$$\mathcal{Y} = \mathcal{T} \times_3 A \in \mathbb{R}^{m \times n \times R}$$

where its elements are defined by

$$y_{ij}^r = \sum_{k=1}^{p} t_{ij}^k a_{rk} \quad \text{where} \quad i = 1, \dots, m, j = 1, \dots, n \text{ and } r = 1, \dots, R.$$

To understand the mode-$n$ product in terms of matrix, consider matrices $A \in \mathbb{R}^{m \times n}$, $B \in \mathbb{R}^{k \times m}$, and $C \in \mathbb{R}^{q \times n}$. By Definition 1.3 we have

$$A \times_1 B = BA \in \mathbb{R}^{k \times n} \text{ and } A \times_2 C = AC^T \in \mathbb{R}^{m \times q}.$$

Thus, the singular value decomposition of matrix $A$ can be written as

$$U \Sigma V^T = (\Sigma \times_1 U) \times_2 V = (\Sigma \times_2 V) \times_1 U.$$

The mode-$n$ product satisfies the following property [8]:

**Property 1.** Let $\mathcal{T}$ be a tensor of order 3 and matrices $A$ and $B$ of convenient sizes. We have for all $r, s = 1, 2, 3$

$$(\mathcal{T} \times_r A) \times_s B = (\mathcal{T} \times_s B) \times_r A = \mathcal{T} \times_r A \times_s B \quad \text{for} \, r \neq s \text{ and} \tag{5}$$

$$(\mathcal{T} \times_r A) \times_r B = \mathcal{T} \times_r (BA) \tag{6}$$

The idea of Bader and Kolda [5] to calculate the product between tensor and vector is to calculate the inner product of each mode-$n$ fiber (column, row, or tube) with the vector. It is not advantageous to treat an $m$-dimensional vector as a matrix $m \times 1$. For example, if we consider a tensor $\mathcal{T} \in \mathbb{R}^{m \times n \times p}$ and a vector $v \in \mathbb{R}^{m \times 1}$, with $m, n, p \neq 1$, by Definition 1.3, the product between $\mathcal{T}$ and $v$ is not well defined, but it is possible to calculate $\mathcal{T} \times_1 v^T$.

**Definition 1.4.** (Contracted product mode-n between tensors and vectors) The contracted product mode-1 between a tensor $\mathcal{T} \in \mathbb{R}^{m \times n \times p}$ and a vector $v \in \mathbb{R}^m$ is the matrix

$$A = \mathcal{T} \overline{\times}_1 v \in \mathbb{R}^{n \times p}$$

where its elements are defined by

$$a_{jk} = \sum_{i=1}^{m} t_{ij}^k v_i \quad \text{where} \quad j = 1, \dots, n \text{ and } k = 1, \dots, p$$

where $v_i$ is the $i$-th component of the vector $v$.

The contracted product mode-2 between a tensor $\mathcal{T} \in \mathbb{R}^{m\times n\times p}$ and a vector $v \in \mathbb{R}^n$ is the matrix

$$A = \mathcal{T}\,\overline{\times}_2 v \in \mathbb{R}^{m\times p}$$

where its elements are defined by

$$a_{ik} = \sum_{j=1}^{n} t_{ij}^{k} v_j \quad \text{where} \quad i = 1, \dots, m \text{ and } k = 1, \dots, p$$

where $v_j$ is the $j$-th component of the vector $v$.

The contracted product mode-3 between a tensor $\mathcal{T} \in \mathbb{R}^{m\times n\times p}$ and a vector $v \in \mathbb{R}^p$ is the matrix

$$A = \mathcal{T}\,\overline{\times}_3 v \in \mathbb{R}^{m\times n}$$

where its elements are defined by

$$a_{ij} = \sum_{k=1}^{p} t_{ij}^{k} v_k \text{ where } i = 1, \dots, m \text{ and } j = 1, \dots, n$$

where $v_k$ is the $k$-th component of the vector $v$.

A caution must be added when calculating the product between matrices and vectors by considering the definitions 1.3 and 1.4. For example, note that if $A \in \mathbb{R}^{m\times n}$, $u \in \mathbb{R}^n$, and $v \in \mathbb{R}^m$, then $A\,\overline{\times}_2 u$ and $A \times_2 u^T$ have the same elements, but

$$A\,\overline{\times}_2 u \neq A \times_2 u^T,$$

because $A\,\overline{\times}_2 u \in \mathbb{R}^m$ (column vector) and $A \times_2 u^T \in \mathbb{R}^{1\times m}$ (row vector). Note that, in relation to the matrix product of elementary algebra, we have

$$Au = A\,\overline{\times}_2 u \tag{7}$$

$$v^T A = A \times_1 v^T \neq A\,\overline{\times}_1 v. \tag{8}$$

In particular, given a tensor $\mathcal{T} \in \mathbb{R}^{n\times m\times m}$ and a vector $v \in \mathbb{R}^m$, by Definition 1.4 together with (8), it follows that $\mathcal{T}\,\overline{\times}_2 v \in \mathbb{R}^{n\times m}$ and

$$(\mathcal{T}\,\overline{\times}_2 v)\,\overline{\times}_2 v = (\mathcal{T}\,\overline{\times}_2 v)v \in \mathbb{R}^n.$$

The contracted product mode-$n$ satisfies the following property [5]:

**Property 2.** Given a tensor $\mathcal{T}$ of order 3 and vectors $u$ and $v$ of convenient sizes, we have for all $r = 1, 2, 3$ and $s = 2, 3$ that

$$(\mathcal{T}\,\overline{\times}_r u)\,\overline{\times}_{s-1} v = (\mathcal{T}\,\overline{\times}_s v)\,\overline{\times}_r u \quad \text{for } r < s.$$

For example, consider a tensor $\mathcal{T} \in \mathbb{R}^{2\times 4\times 3}$, and denote the $k$-th column and the $q$-th row of matrix $A$ by $\mathrm{col}_k(A)$ and $\mathrm{row}_q(A)$, respectively. Note that if:

1.$x \in \mathbb{R}^2$, then $\mathcal{T}\,\overline{\times}_1 x \in \mathbb{R}^{4\times 3}$ and

$$\text{col}_k(\mathcal{T}\overline{\times}_1 x) = \begin{pmatrix} a_{1k} \\ a_{2k} \\ a_{3k} \\ a_{4k} \end{pmatrix} = \begin{pmatrix} t_{11}^k & t_{21}^k \\ t_{12}^k & t_{22}^k \\ t_{13}^k & t_{23}^k \\ t_{14}^k & t_{24}^k \end{pmatrix} \begin{pmatrix} x_1 \\ x_2 \end{pmatrix} = \left(\mathcal{T}^{::k}\right)^T x \text{ and}$$

$$\text{row}_j(\mathcal{T}\overline{\times}_1 x) = \begin{pmatrix} a_{j1} & a_{j2} & a_{j3} \end{pmatrix} = \begin{pmatrix} x_1 & x_2 \end{pmatrix} \begin{pmatrix} t_{1j}^1 & t_{1j}^2 & t_{11}^3 \\ t_{2j}^1 & t_{2j}^2 & t_{21}^3 \end{pmatrix} = x^T \mathcal{T}^{:j:}$$

2.$x \in \mathbb{R}^4$, then $\mathcal{T}\overline{\times}_2 x \in \mathbb{R}^{2\times 3}$ and

$$\text{col}_k(\mathcal{T}\overline{\times}_2 x) = \begin{pmatrix} a_{1k} \\ a_{2k} \end{pmatrix} = \begin{pmatrix} t_{11}^k & t_{12}^k & t_{13}^k & t_{14}^k \\ t_{21}^k & t_{22}^k & t_{23}^k & t_{24}^k \end{pmatrix} \begin{pmatrix} x_1 \\ x_2 \\ x_3 \\ x_4 \end{pmatrix} = \left(\mathcal{T}^{::k}\right) x \text{ and}$$

$$\text{row}_i(\mathcal{T}\overline{\times}_2 x) = \begin{pmatrix} a_{i1} & a_{i2} & a_{i3} \end{pmatrix} = \begin{pmatrix} x_1 & x_2 & x_3 & x_4 \end{pmatrix} \begin{pmatrix} t_{i1}^1 & t_{i1}^2 & t_{i1}^3 \\ t_{i2}^1 & t_{i2}^2 & t_{i2}^3 \\ t_{i3}^1 & t_{i3}^2 & t_{i3}^3 \\ t_{i4}^1 & t_{i4}^2 & t_{i4}^3 \end{pmatrix} = x^T \mathcal{T}^{i::}$$

3.$x \in \mathbb{R}^3$, then $\mathcal{T}\overline{\times}_3 x \in \mathbb{R}^{2\times 4}$ and

$$\text{col}_j(\mathcal{T}\overline{\times}_3 x) = \begin{pmatrix} a_{1j} \\ a_{2j} \end{pmatrix} = \begin{pmatrix} t_{1j}^1 & t_{1j}^2 & t_{1j}^3 \\ t_{2j}^1 & t_{2j}^2 & t_{2j}^3 \end{pmatrix} \begin{pmatrix} x_1 \\ x_2 \\ x_3 \end{pmatrix} = \left(\mathcal{T}^{:j:}\right) x \text{ and}$$

$$\text{row}_i(\mathcal{T}\overline{\times}_3 x) = \begin{pmatrix} a_{i1} & a_{i2} & a_{i3} \end{pmatrix} = \begin{pmatrix} x_1 & x_2 & x_3 \end{pmatrix} \begin{pmatrix} t_{i1}^1 & t_{i2}^1 & t_{i3}^1 & t_{i4}^1 \\ t_{i1}^2 & t_{i2}^2 & t_{i3}^2 & t_{i4}^2 \\ t_{i1}^3 & t_{i2}^3 & t_{i3}^3 & t_{i4}^3 \end{pmatrix} = x^T \left(\mathcal{T}^{i::}\right)^T$$

This example can be easily generalized to arbitrary dimensions. In particular, for a tensor $\mathcal{T} \in \mathbb{R}^{m\times n\times n}$ and a vector $x \in \mathbb{R}^n$, we have

$$\text{row}_i(\mathcal{T}\overline{\times}_2 x) = x^T \mathcal{T}^{i::} \tag{9}$$

$$\text{row}_i(\mathcal{T}\overline{\times}_3 x) = x^T \left(\mathcal{T}^{i::}\right)^T \tag{10}$$

**Lemma 1.5.** Let $\mathcal{T} \in \mathbb{R}^{n\times n\times n}$ be a tensor. If $\mathcal{T}^{i::}$ is a symmetric matrix for all $i = 1, \dots, n$, then

$$(\mathcal{T}\overline{\times}_2 u)v = (\mathcal{T}\overline{\times}_2 v)u$$

for all $u, v \in \mathbb{R}^n$.

*Proof.* By Property 2, it follows that $(\mathcal{T}\overline{\times}_2 u)v = (\mathcal{T}\overline{\times}_3 v)u$. By (10), (11), and the symmetry of $\mathcal{T}^{i::}$, we have $\mathcal{T}\overline{\times}_3 v = \mathcal{T}\overline{\times}_2 v$. □

## 3. Space of bilinear applications

In this section, we define bilinear applications on finite dimensional vector spaces, in order to relate them to the second derivative of a two-differentiable application, as well as a tensor of order 3.

**Definition 1.6.** Let $U, V, W$ be vector spaces. An application $f : U \times V \to W$ is a bilinear application if:

i. $f(\lambda u_1 + u_2, v) = \lambda f(u_1, v) + f(u_2, v)$ for all $\lambda \in \mathbb{R}$, $u_1, u_2 \in U$, and $v \in V$.

ii. $f(u, \lambda v_1 + v_2) = \lambda f(u, v_1) + f(u, v_2)$ for all $\lambda \in \mathbb{R}$, $u \in U$, and $v_1, v_2 \in V$.

In other words, an application $f : U \times V \to W$ is a bilinear application if it is linear in each of the variables when the other variable is fixed. We denote by $\mathcal{B}(U \times V, W)$ the set of all bilinear applications of $U \times V$ in $W$. In particular, if $U = V$ and $W = \mathbb{R}$ in Definition 1.6, then $f : U \times U \to \mathbb{R}$ is a bilinear form in which we are used to quadratic forms, for example.

A simple example of bilinear application is the function $f : U \times V \to \mathbb{R}$ defined by

$$f(u, v) = h(u)g(v), \tag{11}$$

with $h \in U^*$ and $g \in V^*$, where $U^*$ denotes the dual space to $U$. In fact, we have for all $\lambda \in \mathbb{R}, u_1, u_2 \in U$ and $v \in V$ such that

$$f(\lambda u_1 + u_2, v) = h(\lambda u_1 + u_2)g(v) = (\lambda h(u_1) + h(u_2))g(v) = \lambda f(u_1, v) + f(u_2, v).$$

Similarly, it is easy to see that $f(u, \lambda v_1 + v_2) = \lambda f(u, v_1) + f(u, v_2)$ for all $\lambda \in \mathbb{R}, u \in U$ and $v_1, v_2 \in V$.

The next theorem ensures that a bilinear application $f : U \times V \to W$ is well defined when the image of $f$ applied in the bases elements of $U$ and $V$ is known.

**Theorem 1.7.** Let $U$, $V$, and $W$ be vector spaces; $\{u_1, \dots, u_m\}$ and $\{v_1, \dots, v_n\}$ bases of the $U$ and $V$, respectively; and $\{w_{ij} | i = 1, \dots, m \text{ and } j = 1, \dots, n\}$ a subset of $W$. Then, there exists an only bilinear application $f : U \times V \to W$ such that $f(u_i, v_j) = w_{ij}$.

*Proof.* Let $u = \sum_{i=1}^{m} \alpha_i u_i$ and $v = \sum_{j=1}^{n} \beta_j v_j$ be arbitrary elements of $U$ and $V$, respectively. We defined an application $f : U \times V \to W$ by

$$f(u, v) = \sum_{i=1}^{m} \sum_{j=1}^{n} \alpha_i \beta_j w_{ij}.$$

It is easy to see that $f$ is a bilinear application and $f(u_i, v_j) = w_{ij}$. Such an application is unique because if $g$ is another bilinear application satisfying $g(u_i, v_j) = w_{ij}$, then

$$g(u, v) = g\left(\sum_{i=1}^{m} \alpha_i u_i, \sum_{j=1}^{n} \beta_j v_j\right) = \sum_{i=1}^{m} \sum_{j=1}^{n} \alpha_i \beta_j g(u_i, v_j) = \tag{12}$$

$$= \sum_{i=1}^{m} \sum_{j=1}^{n} \alpha_i \beta_j w_{ij} = f(u, v). \tag{13}$$

Therefore $g = f$. □

The following theorem guarantees the isomorphism between space of bilinear applications and space of tensor of order 3.

**Theorem 1.8.** Let $U$, $V$, and $W$ be vector spaces with dimensions $n, p$, and $m$, respectively. Then, the space $\mathcal{B}(U \times V, W)$ has dimension $mnp$.

*Proof*. The idea of the proof is to exhibit a basis for space $\mathcal{B}(U \times V, W)$. For this, let $\{w_1, \dots, w_m\}$, $\{u_1, \dots, u_n\}$, and $\{v_1, \dots, v_p\}$ be bases of the $W$, $U$, and $V$, respectively. For each triple $(i, j, k)$, with $i = 1, \dots, m, j = 1, \dots, n$, and $k = 1, \dots, p$, we define a bilinear application $f_{ij}^k : U \times V \to W$ such that

$$f_{ij}^k(u_r, v_s) = \begin{cases} w_i & \text{if } r = j \text{ and } s = k \\ 0 & \text{if } r \neq j \text{ or } s \neq k. \end{cases} \tag{14}$$

Theorem 1.7 ensures the existence of the $f_{ij}^k$. We will then show that the set

$$A = \left\{ f_{ij}^k | i = 1, \dots, m, j = 1, \dots, n \text{ and } k = 1, \dots, p \right\}$$

is a basis of the space $\mathcal{B}(U \times V, W)$. Let $f \in \mathcal{B}(U \times V, W)$. We note in passing that

$$f(u_r, v_s) = \sum_{i=1}^{m} a_{ir}^s w_i \tag{15}$$

for all $r = 1, \dots, n$ and $s = 1, \dots, p$. Consider the bilinear application

$$g = \sum_{i=1}^{m} \sum_{j=1}^{n} \sum_{k=1}^{p} a_{ij}^k f_{ij}^k.$$

Our goal is to show that $g = f$. In particular, we have

$$g(u_r, v_s) = \sum_{i=1}^{m} \sum_{j=1}^{n} \sum_{k=1}^{p} a_{ij}^k f_{ij}^k(u_r, v_s) = \sum_{i=1}^{m} a_{ir}^s w_i = f(u_r, v_s)$$

for all $r = 1, \dots, n$ and $s = 1, \dots, p$. Therefore $g = f$. The set $A$ is linearly independent, because if

$$\sum_{i=1}^{m} \sum_{j=1}^{n} \sum_{k=1}^{p} a_{ij}^k f_{ij}^k = 0,$$

then

$$0 = \sum_{k=1}^{p} \sum_{i=1}^{m} \sum_{j=1}^{n} a_{ij}^k f_{ij}^k(u_r, v_s) = \sum_{i=1}^{m} a_{ir}^s w_i.$$

Since $\{w_1, \dots, w_m\}$ is a basis of $W$, it follows that $a_{ir}^s = 0$ for all $i = 1, \dots, m, r = 1, \dots, n$, and $k = 1, \dots, p$. □

In particular, if the dimensions of the vector spaces $U$ and $V$ are $m$ and $n$, respectively, then the vector space $\mathcal{B}(U \times V, \mathbb{R})$ has dimension $mn$. Now, as two vector spaces of the same finite dimension are isomorphic [16], there exists a matrix

$m \times n$ associated with each $f \in \mathcal{B}(U \times V, \mathbb{R})$. By considering $B = \{u_1, \dots, u_m\}$ and $C = \{v_1, \dots, v_n\}$ bases of $U$ and $V$, respectively, and if $u = \sum_{i=1}^{m} \alpha_i u_i$ and $v = \sum_{j=1}^{n} \beta_j v_j$, then by doing $f(u_i, v_j) = a_{ij}$ for all $i = 1, \dots, m$ and $j = 1, \dots, n$, we have

$$f(u, v) = \sum_{i=1}^{m} \sum_{j=1}^{n} \alpha_i a_{ij} \beta_j,$$

which in matrix form is $f(u, v) = [u]_B^T A [v]_C$, where $A = (a_{ij})$ and $[v]_C$ denote the vector components $v$ in the basis $C$. Hence follows the next definition:

**Definition 1.9.** Let $U$ and $V$ be vector spaces of finite dimension and ordered bases $B = \{u_1, \dots, u_m\} \subset U$ and $C = \{v_1, \dots, v_n\} \subset V$. We define, for each $f \in \mathcal{B}(U \times V, \mathbb{R})$, the matrix $A = (a_{ij}) \in \mathbb{R}^{m \times n}$ of the $f$ relative to the ordered bases $B$ and $C$, whose elements are given by $a_{ij} = f(u_i, v_j)$ with $i = 1, \dots, m$ and $j = 1, \dots, n$.

Consider now the space $\mathcal{B}(\mathbb{R}^m \times \mathbb{R}^n, \mathbb{R}^p)$ and the canonical bases $\{e_1, \dots, e_m\}$, $\{\bar{e}_1, \dots, \bar{e}_n\}$, $\{\hat{e}_1, \dots, \hat{e}_p\}$ of the $\mathbb{R}^m$, $\mathbb{R}^n$, and $\mathbb{R}^p$, respectively. Consider $f \in \mathcal{B}(\mathbb{R}^m \times \mathbb{R}^n, \mathbb{R}^p)$. For all $u \in \mathbb{R}^m$ and $v \in \mathbb{R}^n$, we have

$$f(u, v) = \sum_{j=1}^{m} \sum_{k=1}^{n} u_j v_k f(e_j, \bar{e}_k)$$

where $u_j$ and $v_k$ are the components of the $u$ and $v$ in the canonical bases of $\mathbb{R}^m$ and $\mathbb{R}^n$, respectively. Denote the $i$-th component of the $f$ by $f_i$. Note that $f_i \in \mathcal{B}(\mathbb{R}^m \times \mathbb{R}^n, \mathbb{R})$. So for each $i = 1, \dots, p$, we have

$$f_i(u, v) = \sum_{j=1}^{m} \sum_{k=1}^{n} u_j v_k f_i(e_j, \bar{e}_k).$$

By Definition 1.9, the matrix of the $f_i$ in relation to the canonical bases is the matrix

$$A_i = \left(t_{ij}^k\right) \in \mathbb{R}^{m \times n},$$

where $t_{ij}^k = f_i(e_j, \bar{e}_k)$. So, we can write

$$f_i(u, v) = u^T A_i v.$$

In general, we can define $p$ matrices $m \times n$ as a tensor $\mathcal{T} \in \mathbb{R}^{p \times m \times n}$; this means that the $p$ matrices can be seen as the horizontal slices of the tensor $\mathcal{T}$. We note in passing that we can write $f(u, v)$ as a product between tensor $\mathcal{T}$ and vectors $u$ and $v$, that is,

$$f(u, v) = \begin{pmatrix} u^T A_1 v \\ u^T A_2 v \\ \vdots \\ u^T A_p v \end{pmatrix} = (\mathcal{T} \overline{\times}_2 u) v. \tag{16}$$

Thus, we can generalize Definition 1.9 as follows:

**Definition 1.10.** Let $U$ and $V$ be finite dimension vector spaces. For fixed bases $B = \{u_1, \dots, u_m\}$ and $C = \{v_1, \dots, v_n\}$ of the $U$ and $V$, respectively, we define, for each $f \in \mathcal{B}(U \times V, \mathbb{R}^p)$, the tensor $\mathcal{T} = \left(t_{ij}^k\right) \in \mathbb{R}^{p\times m\times n}$ of the $f$ relative to the ordered bases $B$ and $C$, whose elements are given by $t_{ij}^k = f_i(u_j, v_k)$ where $f_i$ is the i-th component of the $f$, that is, $f_i \in \mathcal{B}(U \times V, \mathbb{R})$, with $i = 1, \dots, p, j = 1, \dots, m$, and $k = 1, \dots, n$.

## 4. Differentiability

Let $U$ be an open subset of $\mathbb{R}^m$ and $F : U \subset \mathbb{R}^m \to \mathbb{R}^n$ a differentiable application throughout $U$ and $a \in U$. Denote $\mathcal{L}\ (\mathbb{R}^m, \mathbb{R}^n)$ the set of all linear applications of $\mathbb{R}^m$ in $\mathbb{R}^n$. When $F' : U \subset \mathbb{R}^m \to \mathcal{L}\ (\mathbb{R}^m, \mathbb{R}^n)$ is differentiable in $a \in U$, we say that the application $F$ is twice differentiable in $a \in U$ and then the linear transformation $F''(a) \in \mathcal{L}(\mathbb{R}^m, \mathcal{L}\ (\mathbb{R}^m, \mathbb{R}^n))$ is the second derivative of $F$ in $a \in U$.

The norm of $F''(a)$ is naturally defined. For any $h \in \mathbb{R}^m$, it follows that

$$\left\|F''(a)h\right\| = \sup_{\|k\|=1} \left\{\left\|F''(a)hk\right\| \text{ com } k \in \mathbb{R}^m\right\}$$

and then

$$\left\|F''(a)\right\| = \sup_{\|h\|=1} \left\|F''(a)h\right\| = \sup_{\|h\|=1} \sup_{\|k\|=1} \left\|F''(a)hk\right\|.$$

An important observation with respect to Theorem 1.8 is that the spaces $\mathcal{L}(\mathbb{R}^m, \mathcal{L}(\mathbb{R}^m, \mathbb{R}^n))$ and $\mathcal{B}(\mathbb{R}^m \times \mathbb{R}^m, \mathbb{R}^n)$ are isomorphic. This means that $F''(a)$ is a bilinear application belonging to space $\mathcal{B}(\mathbb{R}^m \times \mathbb{R}^m, \mathbb{R}^n)$. Such isomorphism can be found in classical analysis books [17, 18]. On the other hand, by the same theorem, the space of bilinear applications $\mathcal{B}(\mathbb{R}^m \times \mathbb{R}^m, \mathbb{R}^n)$ and space of tensor $\mathbb{R}^{n\times m\times m}$ are also isomorphic.

In many practical applications, such as algorithm implementations, the second derivative $F''(a)$ may be implemented as a tensor belonging to space $\mathbb{R}^{n\times m\times m}$. The question now is how the tensor elements are formed. For this, consider the application $A : \mathbb{R} \to \mathbb{R}^{n\times m}$ and $\alpha \in \mathbb{R}$. We have $A(\alpha)$ as a matrix with $n$ rows and $m$ columns. Its elements are denoted by $a_{ij}(\alpha)$ where $a_{ij}$ are components functions of $A$ with $i = 1, \dots, n$ and $j = 1, \dots, m$. Case $a_{ij} : \mathbb{R} \to \mathbb{R}$ is differentiable in $\alpha$ for all $i = 1, \dots, n$ and $j = 1, \dots, m$; the derivative of $A$ in $\alpha$ is the matrix

$$A'(\alpha) = \left(a'_{ij}(\alpha)\right) \in \mathbb{R}^{n\times m}. \tag{17}$$

The definition of the derivative of $A(\alpha)$ (17) is a classical definition. We refer to [19] for details.

In the sense of generalizing (17), consider now $A : U \subset \mathbb{R}^p \to \mathbb{R}^{n\times m}$ a differentiable application in $u \in U$ with component function $a_{ij} : \mathbb{R}^p \to \mathbb{R}$ with $i = 1, \dots, n$ and $j = 1, \dots, m$. When $a_{ij}$ is differentiable in $u$ for all $i = 1, \dots, n$ and $j = 1, \dots, m$, we defined the derivative of $A$ in $u$ as the tensor

$$A'(u) = \left(\nabla a_{ij}(u)\right) \in \mathbb{R}^{n\times m\times p}. \tag{18}$$

Note that in fact (18) is a generalization of (17). With fixed $i$ and $j$, $\nabla a_{ij}(u)$ is a tube fiber of the tensor $A'(u)$, whose elements are

$$A'(u)_{ij}^{k} = \frac{\partial a_{ij}}{\partial x_k}(u) \tag{19}$$

for all $k = 1, \dots, p$.

For example, consider an application $F : U \subset \mathbb{R}^2 \to \mathbb{R}^3$ twice differentiable in $a \in U$ where $U$ is an open set. The Jacobian matrix of $F$ in $a$ is given by

$$J_F(a) = \begin{pmatrix} \nabla f_1(a)^T \\ \nabla f_2(a)^T \\ \nabla f_3(a)^T \end{pmatrix} = \begin{pmatrix} \frac{\partial f_1}{\partial x_1}(a) & \frac{\partial f_1}{\partial x_2}(a) \\ \frac{\partial f_2}{\partial x_1}(a) & \frac{\partial f_2}{\partial x_2}(a) \\ \frac{\partial f_3}{\partial x_1}(a) & \frac{\partial f_3}{\partial x_2}(a) \end{pmatrix}$$

and its derivative is, by (18), the tensor

$$J_{F'}(a) = \mathcal{T}_F(a) = \left( \nabla \frac{\partial f_i}{\partial x_j}(a) \right) \in \mathbb{R}^{3\times2\times2} \tag{20}$$

where, by (19), its elements are described as

$$t_{ij}^{k} = \frac{\partial^2 f_i}{\partial x_k \partial x_j}(a).$$

With fixed $i$, it is easy to see that the $i$-th horizontal slice of the $\mathcal{T}_F(a)$ is the Hessian matrix $\nabla^2 f_i(a)$ defined by

$$\nabla^2 f_i(a) = \mathcal{T}_F(a)^{i::} = \begin{pmatrix} \frac{\partial^2 f_i}{\partial x_1 \partial x_1}(a) & \frac{\partial^2 f_i}{\partial x_1 \partial x_2}(a) \\ \frac{\partial^2 f_i}{\partial x_2 \partial x_1}(a) & \frac{\partial^2 f_i}{\partial x_2 \partial x_2}(a) \end{pmatrix}. \tag{21}$$

We note in passing that any column of the matrix $\nabla^2 f_i(x)$ is a row fiber of the $i$-th horizontal slice.

As mentioned in the introduction, some numerical methods need to calculate the product between tensor $\mathcal{T}_F(a)$ and vectors in $\mathbb{R}^2$.

From Definition 1.4, it is possible to calculate the contracted product mode-2 and mode-3. As Hessian matrices are symmetrical, given $v \in \mathbb{R}^2$, by Lemma 1.5 together with (10) and (11), we have

$$\mathcal{T}_F(a)\overline{\times}_3 v = \mathcal{T}_F(a)\overline{\times}_2 v = \begin{pmatrix} \mathrm{row}_1(\mathcal{T}_F(a)\overline{\times}_2 v) \\ \mathrm{row}_2(\mathcal{T}_F(a)\overline{\times}_2 v) \\ \mathrm{row}_3(\mathcal{T}_F(a)\overline{\times}_2 v) \end{pmatrix} = \begin{pmatrix} v^T \nabla^2 f_1(a) \\ v^T \nabla^2 f_2(a) \\ v^T \nabla^2 f_3(a) \end{pmatrix} \in \mathbb{R}^{3\times2}$$

and consequently it follows that

$$(\mathcal{T}_F(a)\overline{\times}_2 v)u = \begin{pmatrix} v^T\nabla^2 f_1(a)u \\ v^T\nabla^2 f_2(a)u \\ v^T\nabla^2 f_3(a)u \end{pmatrix} \in \mathbb{R}^3 \tag{22}$$

for all $u, v \in \mathbb{R}^2$.

This means that the tensor $\mathcal{T}_F(a)$ defined by (20) is the associate tensor to bilinear application $F''(a)$, in relation to canonical basis of $\mathbb{R}^2$, by means of Definition 1.10. Without loss of generality, we have

$$\mathcal{T}_F(a)\overline{\times}_3 v = \mathcal{T}_F(a)\overline{\times}_2 v = \mathcal{T}_F(a)v$$

and by means of Lemma 1.5, it follows that

$$(\mathcal{T}_F(a)u)v = (\mathcal{T}_F(a)v)u = \mathcal{T}_F(a)vu.$$

To finish, we consider the following particular case. We know that the $k$-th column of Jacobian $J_F(x)$ is equal to product $J_F(x)e_k$, where $e_k$ is the $k$-th canonical vector of $\mathbb{R}^n$. It is worth noting what the slice of the matrix $\mathcal{T}_F(x)e_k$ is. By definition, we have

$$\mathcal{T}_F(x)e_k = \begin{pmatrix} e_k^T\nabla^2 f_1(x) \\ e_k^T\nabla^2 f_2(x) \\ \vdots \\ e_k^T\nabla^2 f_n(x) \end{pmatrix} = \begin{pmatrix} \text{row}_k\nabla^2 f_1(x) \\ \text{row}_k\nabla^2 f_2(x) \\ \vdots \\ \text{row}_k\nabla^2 f_n(x) \end{pmatrix}$$

Given that $\text{row}_k\nabla^2 f_i(x)$ is the $k$-th tube fiber of $i$-th horizontal slice, we have $\mathcal{T}_F(x)e_k$ as the $k$-th lateral slice, or, by symmetry of Hessians, it is the transpose of $k$-th frontal slice. In short, for the twice differentiable application $F : U \subset \mathbb{R}^n \to \mathbb{R}^m$, we have $\mathcal{T}_F(x) \in \mathbb{R}^{m\times n\times n}$ where the $m$ horizontal slices are the Hessians $\nabla^2 f_i(x)$, with $i = 1, \ldots, m$ and the $n$ lateral and frontal slices obtained by the following product $\mathcal{T}_F(x)e_k$, with $k = 1, \ldots, n$.

## 5. Conclusions

In this text, we have shown some properties of tensors, in particular those of order 3. In addition, we have approached bilinear applications, and we have shown the isomorphism between space of bilinear applications and of tensor of order 3. As mentioned in the introduction, to solve a nonlinear system, some numerical methods use tensors, either in the iterative scheme or in the proof of theorems. For this reason, we have written a section on differentiability of applications by showing how to calculate the product between tensor of second derivatives and vectors.

## Author details

Rodrigo Garcia Eustaquio
Department of Mathematics, Federal Technological University of Paraná, Curitiba, PR, Brazil

*Address all correspondence to: eustaquio@utfpr.edu.br; rodrigogeustaquio@gmail.com

## References

[1] Hernández MA, Gutiérrez JM. A family of Chebyshev-Halley type methods in Banach spaces. Bulletin of the Australian Mathematical Society. 1997;**55**:113-130

[2] Ehle GP, Schwetlick H. Discretized Euler-Chebyshev multistep methods. SIAM Journal on Numerical Analysis. 1976;**13**(3):432-447

[3] Eustaquio RG, Ribeiro AA, Dumett MA. A new class of root-finding methods in $\mathbb{R}^n$: The inexact tensor-free Chebyshev-Halley class. Computational and Applied Mathematics. 2018;**37**: 6654-6675

[4] Traub JF. Iterative Methods for the Solution of Equations. Prentice-Hall Series in Automatic Computation. Englewood Cliffs, NJ: Prentice-Hall; 1964

[5] Bader BW, Kolda TG. Algorithm 862: MATLAB tensor classes for fast algorithm prototyping. ACM Transactions on Mathematical Software. 2006;**32**(4):635-653. DOI: 10.1145/1186785.1186794

[6] Cichocki A, Zdunek R, Phan AH, Amari S. Nonnegative Matrix and Tensor Factorizations: Applications to Exploratory Multiway Data Analysis and Blind Source Separation. John Wiley Sons, Ltd; 2009

[7] Kolda TG, Bader BW. Tensor decompositions and applications. SIAM Review. 2009;**51**(3):455-500

[8] De Lathauwer L, De Moor B, VandeWalle J. A multilinear singular value decomposition. SIAM Journal on Matrix Analysis and Applications. 2000; **21**:1253-1278

[9] Smilde A, Bro R, Geladi P. Multi-Way Analysis: Applications in the Chemical Sciences. Wiley; 2004

[10] Chen B, Petropulu A, De Lathauwer L. Blind identification of convolutive MIMO systems with 3 sources and 2 sensors. Applied Signal Processing. 2002;**5**:487-496. Special Issue: Space-time Coding and Its Applications—Part II. Available from: http://publi-etis.ensea.fr/2002/CPD02a

[11] Golub GH, Kolda TG, Nagy JG, Van Loan CF. Workshop on Tensor Decompositions. Palo Alto, California: American Institute of Mathematics; 2004. Available from http://www.aimath.org/WWN/tensordecomp/

[12] De Lathauwer L, Comon P. Workshop on Tensor Decompositions and Applications. Marseille, France; 2005. Available from: http://www.etis.ensea.fr/wtda/

[13] Bader BW, Kolda TG. Efficient MATLAB Computations with Sparse and Factored Tensors. Albuquerque, NM/Livermore, CA: Sandia National Laboratories; 2006. SAND2006-7592. Available from: http://www.prod.sandia.gov/cgi-bin/techlib/access-control.pl/2006/067592.pdf

[14] Ishteva M. Numerical methods for the best low multilinear rank approximation of higher-order tensors [PhD thesis]. Belgium: Faculty of Engineering, Katholieke Universiteit Leuven; 2009

[15] Kiers HAL. Towards a standardized notation and terminology in multiway analysis. Journal of Chemometrics. 2000;**14**:105-122

[16] Coelho FU, Loureno ML. Um Curso de lgebra Linear. So Paulo, Brasil: Editora da Universidade de So Paulo; 2007

[17] Lima EL. Anlise no Espao $\mathbb{R}^n$. So Paulo: Editora Universidade de Braslia; 1970

[18] Lima EL. Curso de Anlise, Volume 2. Rio de Janeiro, Brasil: IMPA; 1981

[19] Golub GH, Van Loan CF. Matrix Computations. The Johns Hopkins University Press; 1996

Section 2

# Tensors in the Exploring of the Space-Time

Chapter 2

# Kinematic-Energy Measurements of the Torsion Tensor in Space-Time

*Francisco Bulnes, Isaías Martínez, Omar Zamudio and Edgar Navarro*

## Abstract

We consider the relation between the twistor kinematic-energy model of the space-time and the kinematic-energy tensor as the energy-matter tensor studied in relativity theory to obtain the torsion tensor of the space-time. Measurements of the torsion tensor through their energy spectra are obtained for the movement of a particle under certain trajectories (curves whose tangent spaces twist around when they are parallel transported) when crossing an electromagnetic field. We want to give an indicium of the existence of torsion field through the electronic signals produced between the presence of electromagnetic field and the proximity of movement of matter.

**Keywords:** energy-matter tensor, kinematic-energy tensor, movement energy vacuum, torsion tensor, twistor kinematic-energy model

## 1. Introduction

The fundamental problem considered in this chapter is linked with the determination of energy-(space-)time variations that occur in the interaction of movement and matter-energy on a special geometry of movement or movement kinematics. However, we need a background component that permits the measure and detects under the invariance of its fields the change of matter particle spin (as could be in the torsion case [1], considering a quasi-local matter model represented through the gravitational waves of cylindrical type to measure and detect the field torsion). This last, considering only a component of geometrical torsion no vanish, along of a curve of a particle as study object that moves affected by an energy radiation that permits the use of some physical effect like the Hall effect.

The gauging of the torsion system using movement in an external field, which acts on a particle through the deformation space, could be the simplest way to use the dual concepts of twistor frame and spinors. The objective is to demonstrate the existence of the kinematic twistor tensor in a system that detects the torsion and obtains its image by spinors due to the duality, as demonstrated in Ref. [2].

We know the need of an intermediate gauge field to establish experimentally the relation between the kinematic twistor tensor and the energy-matter tensor (this last due to the movement in the space-time) in duality, as determined in Ref. [3].

Likewise, we consider M the space-time as the complex Minkowski model, and we define the kinematic twistor tensor as the obtained of the model in a space region $\Sigma$. Then considering the energy-matter tensor and its image in a two-dimensional surface will be two-surface twistor $\mathbb{T}(S)$. The geometrical evidence of torsion is precisely through this contorted surface.

In other words, the kinematic twistor tensor $A_{\alpha\beta}$ in the radiation energy bath (electromagnetic radiation) from the energy-matter tensor $T_{\alpha\beta}$ will be defined by the interaction of two fields $Z_1^{\alpha}$ and $Z_2^{\alpha}$ that act in $\Sigma$,

$$A_{\alpha\beta}Z_1^{\alpha}Z_2^{\beta} = \int_{\Sigma} T_{\alpha\beta}k^{\alpha}d\sigma^{\beta}, \tag{1}$$

which produces an electrical total charge due to the Gauss divergence theorem on currents $T_{\alpha\beta}k^{\alpha}$,

$$Q[k] = \frac{1}{4\pi G}\int_{\Sigma} R_{\alpha\beta\gamma\delta}f^{\alpha\beta}d\sigma^{\gamma\delta}, \tag{2}$$

This can be identified as the source depending on the killing vector $k^{\alpha}$ of the Minkowski space background model

$$M = S^2 \otimes \mathbb{C}^2 \otimes \mathrm{M}, \tag{3}$$

where M is the space-time of two components

$$\mathrm{M} = S^{+} \oplus S^{-}. \tag{4}$$

Then, its system has a complex set of four-dimensional solution families $(\cong \mathbb{C}^2)$, and the family defines the two-surface twistor space $\mathbb{T}(S)$.

Likewise, we can define the space of kinematic twistor tensor as the space of tensors [2]:

$$(\mathbb{T}(S)\odot\mathbb{T}(S))* = \{A_{\alpha\beta} \in T_2^4(\mathrm{M}) | A_{\alpha\beta}Z^{\alpha}Z^{\beta} = Q[k]\}, \tag{5}$$

Though a gauge field (electromagnetic field as photons) acts on the back-ground radiation of the Minkowski space $M$, and the energy of the matter will be related to this gauge field through the equation

$$j^{\alpha} = T_{\alpha\beta}k^{\alpha}, \tag{6}$$

where $k^{\alpha}$ can represent the density of background radiation, which establishes the curved part of the space (with spherical symmetry) together with $T_{\alpha\beta}$ (see **Figure 1**)

$$Q[k] = \frac{1}{4\pi G}\int_{S^2} T_{\alpha\beta}k^{\alpha}d\sigma^{\beta} \geq \int_{S^2} j^{\alpha}d\sigma^{\beta} \geq 2\pi\chi, \tag{7}$$

The corresponding electromagnetic device generates an electromagnetic radiation bath in a space region, where a movement of mass is detected inside this region, producing variations in the electromagnetic field. If we use a curvature energy sensor [3–5], we will obtain a spectrum in a twistor-spinor frame.

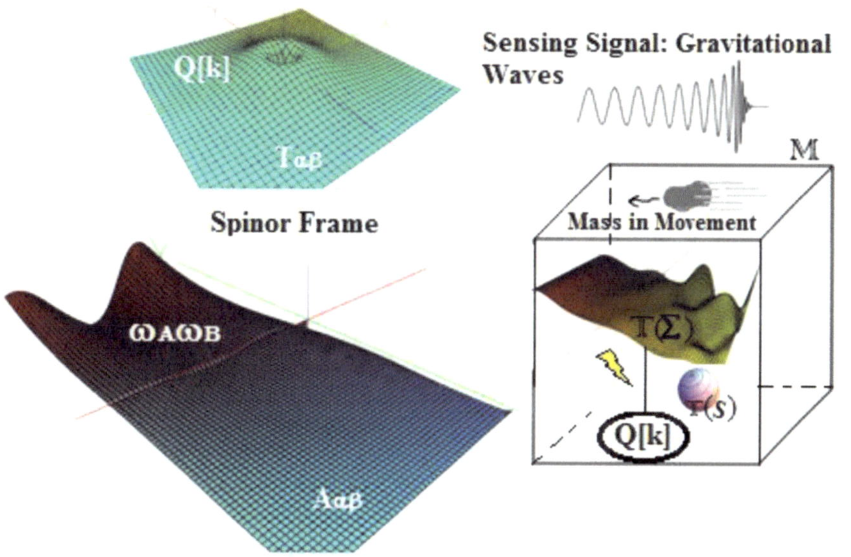


**Figure 1.**
*Supermassive mass movement field + electromagnetic energy field = torsion evidence on the surface of sensing (sphere). How can we construct a tensor whose evidence of torsion can trace the electronic signals that could come from matter and electromagnetic fields of gravitational waves? We need a tensor of invariants of movement identified by invariants in geometry. This is the kinematic twistor tensor $A_{\alpha\beta}$. The space S is the sphere that senses the torsion and transmits its variation at the time to the surface Σ defined by the electromagnetic-matter movement. Two-dimensional model of spinor representation of the kinematic twistor tensor $A_{\alpha\beta}$ is constructed from the sphere.*

Likewise, by the twistor-spinor theory, and by using the duality between the tensors $T_{\alpha\beta}$, and $A_{\alpha\beta}$, we can determine the mechanism of measurement and characterize the geometrical context of the detection of torsion. We define the twistor space as the points set[1]

$$\mathbb{T} = \{Z^{\alpha} | Z^{\alpha} = (\omega^{A}, \pi_{A})\}, \tag{8}$$

for all coordinates systems A and A. We define the twistor infinity tensor $I_{\alpha\beta'}$[2] as the obtained directly of the all space-time whose structure obeys a Minkowski space $M$. Then the surface Σ, which is a 3-dimensional surface is obtained for the twistor fields $Z^{\alpha}$ and $Z^{\alpha}$, that is to say:

$$\Sigma = \Sigma(Z^{\alpha}, Z^{\beta}), \tag{9}$$

which has a metric defined when $\alpha = \beta$ and $Z^{\beta} = \overline{Z^{\beta}}$ (its complex conjugate). Then, in the infinity of the space-time, we have the sequence of mappings:

$$\mathbb{T} \xrightarrow{I^{\alpha\beta}} \mathbb{T}(S) \xrightarrow{I^{\alpha\beta}_{\Sigma}} \mathbb{T}(\Sigma) \tag{10}$$

---

[1] $\omega^{A} : \mathbb{T}* \to \mathbb{T}$, with rule of correspondence on points of the space–time $\pi_{A'} \mapsto ix^{AA'}\pi_{A'}$. Also its dual $\pi_{A'}$ : $\mathbb{T} \to \mathbb{T}*$, with correspondence rule of points of the space–time $\omega^{A} \mapsto -ix^{AA'}\omega^{A}$. Likewise, the corresponding twistor spaces in this case are:

$$\mathbb{T} = \{Z^{\alpha} = (\omega^{A}, \pi_{A'}) | \omega^{A} = ix^{AA'}\pi_{A'}\}, \quad \mathbb{T}* = \{W_{\alpha} = (\pi_{A}, \omega^{A'}) | \omega^{A'} = -ix^{AA'}\pi_{A}\},$$

[2] $I_{\alpha\beta} : \mathbb{T}* \to \mathbb{T}$, with the correspondence rule $W_{\alpha} \mapsto Z^{\alpha} I^{\alpha\beta} W_{\beta}$.

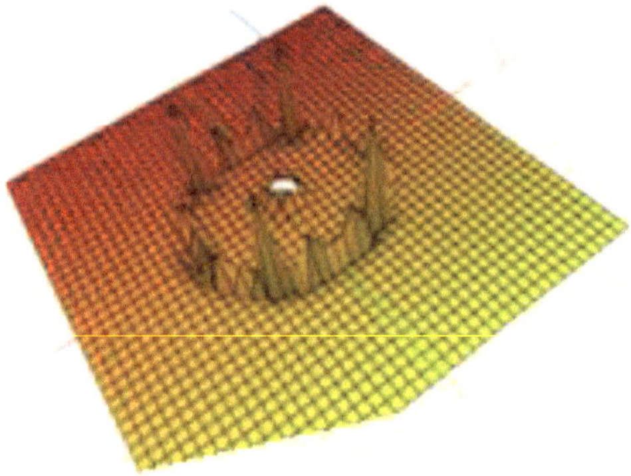

**Figure 2.**
*Kinematic twistor tensor due to the energy-matter tensor perturbation of the supermassive body, which is determined on sphere* S.

whose correspondence rule is given as follows:

$$Z^{\alpha} \mapsto I^{\alpha\beta} S_{\beta\beta'} \overline{Z^{\beta\prime}} \mapsto I^{\alpha\beta} \Sigma_{\beta\beta''} \overline{Z^{\beta\prime}}. \tag{11}$$

We consider the symmetric part of the fields $Z^{\alpha}$ and $Z^{\beta}$, given by the spinors $\omega^{AB}$, which satisfy the valence-2 twistor equation:

$$\nabla^{A}_{A'}\omega^{BC} = -i \in^{A\left(Bk^{C}_{A'}\right)}, \tag{12}$$

which has a solution in a 10-dimensional space. We need limit the space region of our study to spinor waves in a four-dimensional space, that is, on a component of Eq. (3). The solution in the space of Eq. (12) is spanned by spinor fields $\omega^{AB}$ of the form[3]

$$\omega^{AB} = \omega_1{}^{\left(A\omega_2{}^{B}\right)=\omega^{A}\omega^{B}}, \tag{13}$$

where each $\omega_i^{A}$ is a valence-1 twistor, satisfying the equation:

$$\nabla^{A}_{A'}\omega^{B} = -i \in^{AB}\pi_{A'}, \tag{14}$$

We need in all time, for our measurements the conservation condition, which will be given by the equation:

$$\nabla^{\alpha}T^{\alpha\beta} = 0, \tag{15}$$

that is to say, we suppose that the energy-matter is always present in the space and is constant, at least in the space region where is bounded the three-dimensional surface Σ. Likewise, when a supermassive body exists that perturbs the space-time, the energy matter of its tensor can be carried out (see **Figure 2**):

---

[3] Here the spinors product $\omega_1{}^{\left(A\omega_2{}^{B}\right)}$, comes from fields product $Z_1{}^{\left(AZ_2{}^{B}\right)}$, which is a symmetric tensor product, that is to say,

$$Z_1{}^{\left(AZ_2{}^{B}\right)} = Z_1 \otimes_{Symm} Z_2 \in \mathbb{T} \otimes_{Symm} \mathbb{T} = \mathbb{T} \odot \mathbb{T}.$$

$$A_{\alpha\beta} = \frac{1}{16\pi G}\oint_S R_{AB}\omega_\alpha^A\omega_\beta^B, \tag{16}$$

Finally, we can establish the following commutative diagram of twistor space mappings on the gauge and detection mechanism of torsion:

$$\begin{array}{ccccc} \mathbb{T}(\Sigma) & \overset{I^{\alpha\beta}\Sigma_{\beta\beta'}}{\longleftarrow} & \Sigma & \overset{A_{\alpha\beta}}{\longrightarrow} & (\mathbb{T}(S)\odot\mathbb{T}(S))* \\ I_\Sigma^{\alpha\beta}\uparrow & & \uparrow T^{\alpha\beta} & & \uparrow A_{\alpha\beta}Z^\alpha Z^\beta, \\ \mathbb{T}(S) & \overset{I^{\alpha\beta}}{\longleftarrow} & S & \overset{\omega^{AB}}{\longrightarrow} & \mathbb{T}(S)\odot\mathbb{T}(S) \end{array} \tag{17}$$

where $\odot$ is a symmetric tensor product.

## 2. Torsion indicium in gravitational spin waves

In this context, the use of the Einstein-Cartan-Sciama-Kibble theory is important. Likewise, this theory is convenient considering our space-time model as has been defined $M$, and the field experiments considering external fields created through the use of the spin Hall effect and movement of matter in $\Sigma$. We consider the curvature and twistor-spinor framework studied in Refs. [2, 4], where they recover the most important cause of the second curvature.

Likewise, for the curvature tensor $K_{\alpha\beta\gamma\delta}$, we start with the Riemann tensor $R_{\alpha\beta\gamma\delta}$ that appears in the integral (2). Likewise, considering the space-time $M$, a complex Riemannian manifold, we have the conjecture where the indicium of torsion exists [1, 2].

**Conjecture 2.1 (Bulnes F, Rabinovich I).** The curvature in the spinor-twistor framework can be perceived with the appearance of the torsion and the anti-self-dual fields.

*Proof.* [2].

In the previous research of this conjecture [2], it was established that the spinor model of torsion can be written as follows:

$$S_{\alpha\beta}^\gamma = \chi_{AA'}^{CC'} \in_{A'B'} + \tilde{\chi}_{A'B'}^{CC'} \in_{AB}, \tag{18}$$

where it is clear that

$$T_{\alpha\beta}^\gamma = 2S_{\alpha\beta}^\gamma, \tag{19}$$

Then, it is obvious that the torsion tensor can be written as follows:

$$T_{\alpha\beta}^\gamma = 2\left(\chi_{AA'}^{CC'} \in_{A'B'} + \tilde{\chi}_{A'B'}^{CC'} \in_{AB}\right), \tag{20}$$

Considering the spinor equation of torsion (15) in the twistor-spinor framework, we have the transformation in the infinity twistor of the space-time:

$$I_{\alpha\beta} = \pi_\alpha^{A'}\pi_{\beta A'}, \tag{21}$$

and for other transformation of spinor coordinate frame (and derivative), we have:

$$\pi^{A'}(\nabla_{AA'}\pi_{B'}) = \xi_A\pi_{B'} - 2\pi^{A'}\pi^{C'}\chi_{A'B'AC'}, \tag{22}$$

## 3. Curvature energy to torsion

The following results obtained in Ref. [2] are the fundamental principles that are required to gauge and detect the torsion through the tensor $A_{\alpha\beta}$, considering the law transformation to pass from a field $Z^{\alpha}$ to other $Z^{\beta}$ through two coordinate systems $\alpha$ and $\beta$ to transform the surface $\Sigma$:

$$\Sigma_{\alpha\beta} = A_{\alpha\beta} I^{\beta\gamma} \Sigma_{\gamma\alpha'} \tag{23}$$

Then, we enunciate the following theorem.

**Theorem 3.1 (Bulnes F, Stropovsvky Y, Rabinovich I).** We consider the embedding as follows:

$$\sigma : \Sigma \to (\mathbb{T}(S) \otimes \mathbb{T}(S))*, \tag{24}$$

The space $\sigma(\Sigma)$ is smoothly embedded in the twistor space $(\mathbb{T}(S) \otimes \mathbb{T}(S))*$. Then, their curvature energy is given in the interval $M_N \geq A_{\alpha\beta} Z^{\alpha} I^{\beta\gamma} \overline{Z}_{\gamma} \geq 0$.

*Proof*. [2].

We have a source to linearized gravitational field that is explained through kinematics and electrodynamics used in its construction (see **Figure 3**). The linearized Riemann tensor corresponding to the spinor frame has been constructed, considering the components

$$f_{\alpha\beta} = \omega_{AB} \in_{A'B'}, \tag{25}$$

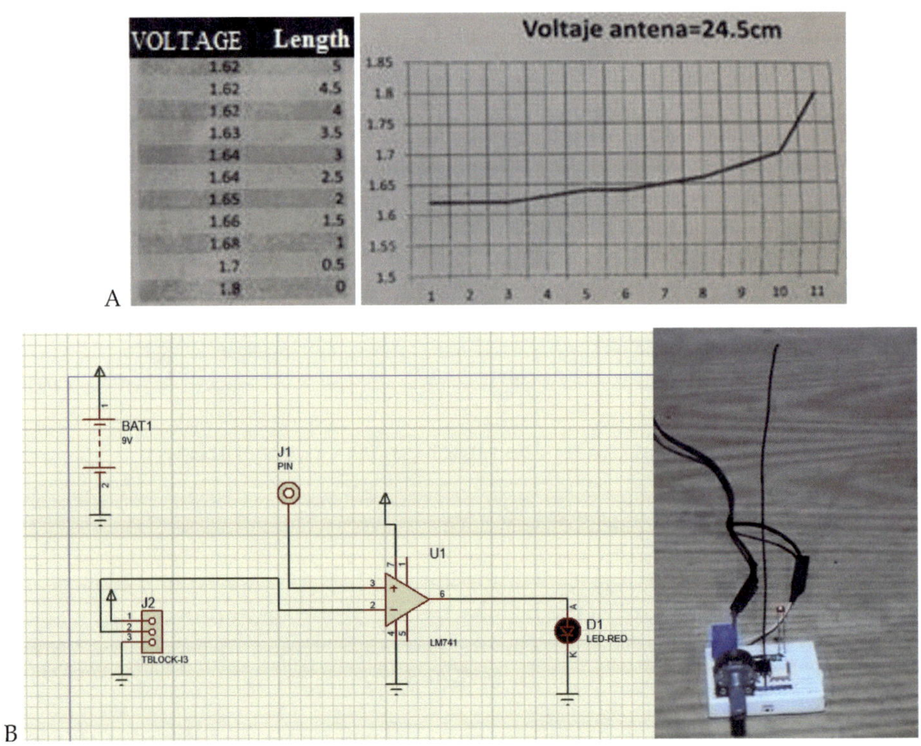

| VOLTAGE | Length |
|---|---|
| 1.62 | 5 |
| 1.62 | 4.5 |
| 1.62 | 4 |
| 1.63 | 3.5 |
| 1.64 | 3 |
| 1.64 | 2.5 |
| 1.65 | 2 |
| 1.66 | 1.5 |
| 1.68 | 1 |
| 1.7 | 0.5 |
| 1.8 | 0 |



**Figure 3.**
*(A) Antenna with voltage feeding of length 24.5 cm. (B) Electronic device of electronic monopole to electromagnetic radiation bath.*

which relates to the spinor field $\omega_{AB}$, with the killing vector $k^{\alpha}$, in the valence-2 twistor equation. We use the divergence theorem when S is a 2-surface in the 3-surface $\Sigma$, which is given as follows:

$$\Sigma = \omega^{A}\overline{\pi}_{A'} + \omega^{A}\overline{\pi}_{A'}, \tag{26}$$

around the source having several censorship conditions designed through dominating energy conditions of curvature that can be used in the electronic experiments.

We have a metrology [5–7] of curvature measured and detected by our curvature sensors, which permitted us to have the curvature in new units obtained under the strong electronic gauging study [3, 7].

Likewise, the energy of the kinematic twistor tensor that will be substantive energy to curvature energy measure in the case of the spinor-twistor framework is given in the energy domain $M_N \geq A_{\alpha\beta}Z^{\alpha}I^{\beta\gamma}\overline{Z}_{\gamma} \geq 0$.

Then, the solution of the quasi-local mass is directly related to the quantity of energy-matter tensor. Likewise, this solution is a function of radius and time as wave pulse, which can be spectrally reproduced in a function $\frac{\sin\omega L}{\omega L}$, under voltage of the electronic device of electromagnetic radiation bath interacting with the proximity of supermassive object or simple mass movement (see **Figure 2**, and **Figure 3 (A)** and **(B)**).

## 4. Electronic experiment demonstration of torsion existence through wave links such as spinors and wave pulses

An electromagnetic field as detector can also be a part of establishing the perturbation in the space-time that must help us to perceive the torsion existence. Likewise, this field as a solution of the Maxwell equations in the spinor-twistor framework (**Figure 4**)[4] complies the integrals:

$$\varphi_{A'B'\ldots L'}(\mathbb{R}^{a}) = \frac{1}{2\pi i}\oint_{Z^{\alpha}\leftrightarrow\mathbb{R}^{a}} \pi_{A'}\cdots\pi_{L'}f(Z)\pi_{F'}d\pi_{F'} \tag{27}$$

and

$$\varphi_{AB\ldots L'}(\mathbb{R}^{a}) = \frac{1}{2\pi i}\oint_{Z^{\alpha}\leftrightarrow\mathbb{R}^{a}} \frac{\partial}{\partial\omega^{A}}\cdots\frac{\partial}{\partial\omega^{L}}f(Z)\pi_{F'}d\pi_{F'} \tag{28}$$

which for the particular case of the determination of $A_{\alpha\beta}$, are the integrals:

$$\begin{aligned} A_{\alpha\beta} &= \frac{1}{16\pi G}\oint_{S} R_{AB}\omega_{\alpha}^{A}\omega_{\beta}^{B} \\ &= \frac{1}{16\pi G}\oint_{S} R_{\alpha\beta\gamma\delta}f^{\alpha\beta}d\sigma^{\gamma\delta}, \\ &= \frac{1}{2\pi i}\frac{i}{8\pi G}\oint_{S}\omega_{\alpha}^{A}d^{2}\omega_{\beta A} \\ &= \frac{1}{16\pi G}\oint_{S}\omega_{\alpha}^{A}d^{2}\omega_{\beta A}, \end{aligned}$$

[4] Here our electromagnetic wave equation can be characterized by the massless field equations:

$$\nabla^{AA'}\varphi_{AB\ldots L} = 0, \quad \nabla^{AA'}\varphi_{A'B'\ldots L'} = 0,$$

which are equivalent to $\Box\varphi = 0$, for zero spin case.

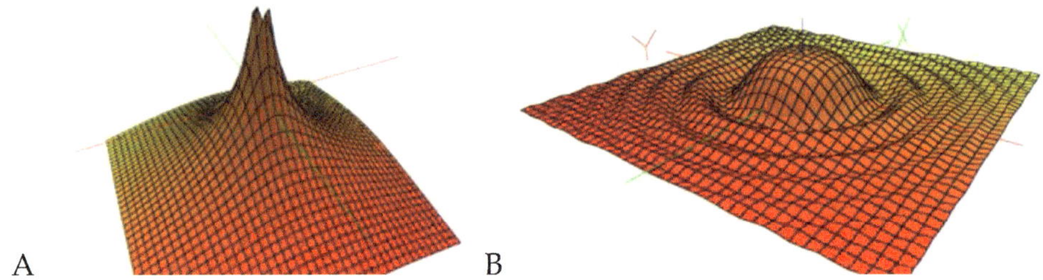


**Figure 4.**
*(A) Two-dimensional surface of charge* Q[k] *in monopole field. (B) Two-dimensional surface of energy-matter tensor* $T^{\alpha\beta}$ *in supermassive body.*

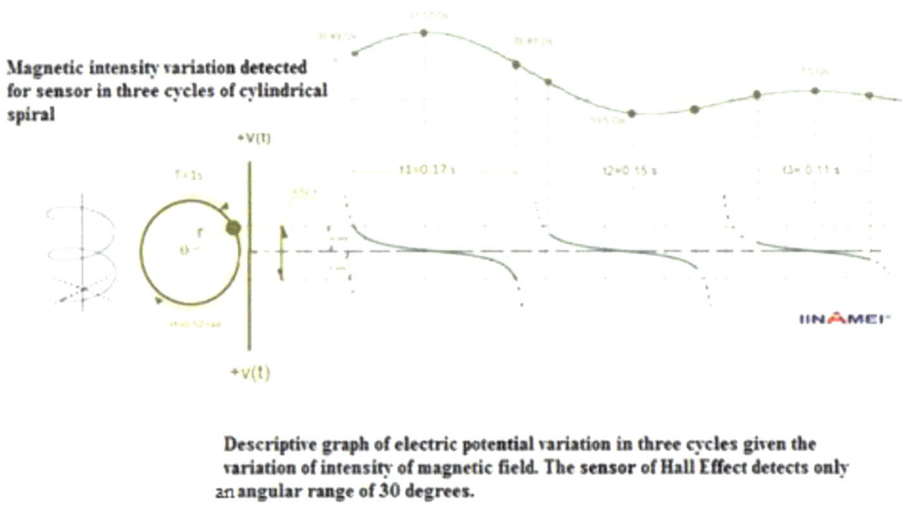


**Figure 5.**
*Dynamic-magnetic system defining the formula* $A_{\alpha\beta}Z^{\alpha}I^{\beta\gamma}\overline{Z}_{\gamma}$..

where it has been applied in the field around the circle used as cycle of the displacement along the three-cylindrical spiral cycles (see **Figure 5**). As discussed in Section 2, the torsion evidence can be obtained with a good approximation (given the limitations of the electronic system) when a complete signal $\frac{\sin\omega L}{\omega L}$ is obtained in each three cycles, where two complete spinors are produced.

The sensor device of magnetic field of Hall effect has detected the boundary whose region is an arco length of 0.045 m (see **Figure 6(A)**). Without this range, there is no detection of field, although it is evident the cyclic subsequent displacements of the magnetic dilaton. This is shown with three curves in the graph of **Figure 5**, with displacement times $t_1, t_2,$ and $t_3$ The electric potential that is generated due to the magnetic field variation is inversely proportional to the magnetic field intensity with base in the relation of 19.4 mV/Gs (**Figure 7**).

In the first half of walk, the magnetic dilaton generates a decreasing potential of 0.52 V, until a minimum of 0.26 V. In the second half of walk, the magnetic dilaton generates an increasing potential of 0.26 V, until a maximum of 0.52 V, when it moves away. For the subsequent cycles, the remoteness of sensor in the trajectory obeys the spiral trajectory of the dynamic system. Both the effect of magnetic dilaton and the dynamics of system define our kinematic twistor tensor $A_{\alpha\beta}$, which can be gauged in a more fine way with a quantum electronic device version of our electronic system used in this experimentation. The tensor of energy mass depends

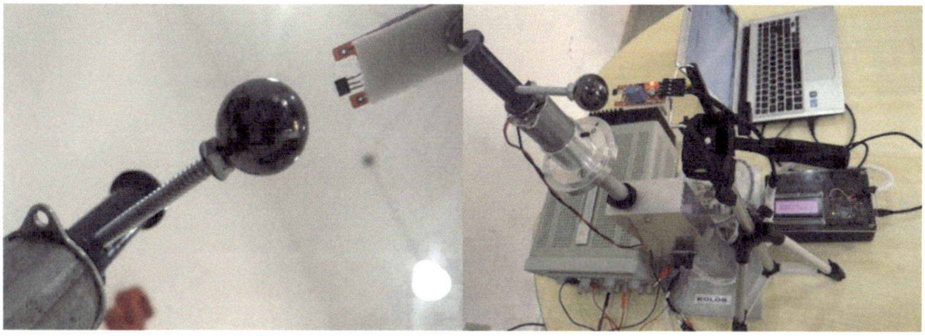

A B

**Figure 6.**
*(A) Magnetic sphere as magnetic dilaton of radius 0.025 m. This dilaton will be used to detect the distortion in the boundary surface Σ, where the interaction happens between the magnetic field of proper dilaton and the gravitational field generated for the mass of the proper dilaton and the mass of the earth. Maximum proximity of sensor is 0.010 m. (B) Rotational dynamic system of radius 0.085, with a reversible vertical displacement of 0.040 m. The sensor used is the Hall effect sensor. The device has an interface system for microcontroller and symmetric variable voltage source.*

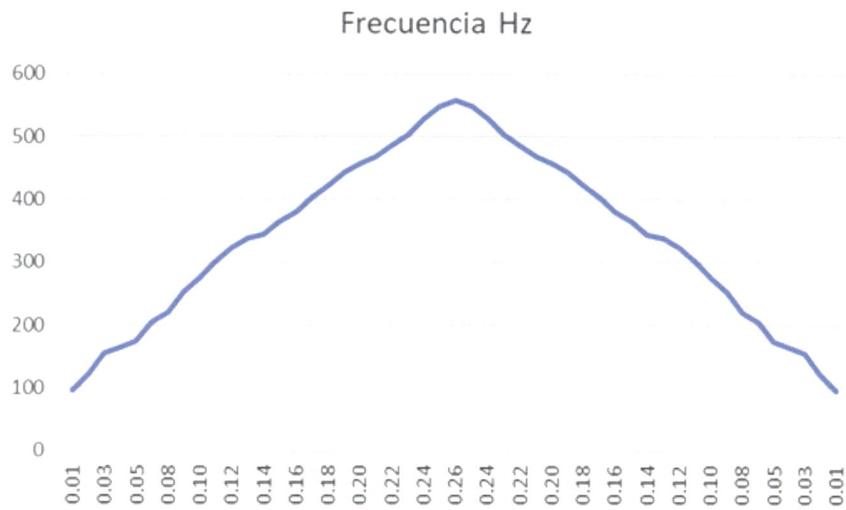


**Figure 7.**
*Frequency in the trajectory of the first cycle.*

on the gravitational field between the dilaton mass and the Earth mass. The coordinate systems A, B, … , L and A′, B′, … , L′ are considered in our inertial reference frames used in the experiment.

The conditioning signal is defined for the continuous variations of the electric potential, which are converted in frequency through the integrated circuit LM331 (see the **Figure 7**). The maximum response (output of frequency) of this device is 10 KHz; therefore, it is developed an electronic circuit to condition the signal and has required lectures. The digital signal obtaining each electric potential variation (0.52–0.26 V, and 0.26–0.52 V) as result of position change of the magnetic dilaton in the space is established. The intention of consider digital signal with pulse width to each respective 26 positions in the space is to do for each pulse a convolution with sinusoidal signal, this to obtain and try with periodic signals to the points study that

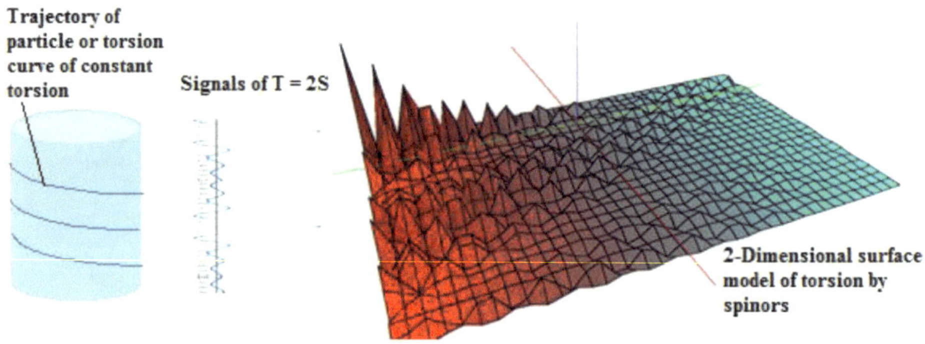


**Figure 8.**
*Two-dimensional model of torsion by spinors.*

determine the curve in a 3-dimensional space in field theory in terms of the signal analysis.

In the first experiment (as described in Section 3), the sphere S has not curved inside the three-dimensional surface $\Sigma$. The electromagnetic field of monopole is fixed and does not produce distortion in the space. Any matter particle complies the spherical symmetry falling in the natural gravitational Earth field.

In the two experiments (in this Section 4), the choose of a magnetic dilaton represented by the ball of certain mass, which is displaced along the cylindrical spiral trajectory, produces a distortion at least in electronic device lectures and in the space, which could be affected for the Earth magnetic field and also for the gravitational field between the dilaton mass and the Earth mass. Summarizing the above, we can consider the following two-dimensional surface model of spinors deduced directly of second experiment verifying some conclusions on the torsion existence and consistence though twistors (see **Figure 8**).

## 5. Conclusions

We can establish different dualities in field theory, geometry, and movement to relate the energy-matter tensor and the kinematic twistor tensor for the torsion study. The torsion is a field observable, which in geometry is a second curvature. From a point of view of the field theory, torsion is an high evidence of the birth gravity and its consequences until our days with the gravitational waves detected from astronomical observatories.

Through of electronics is designed an analogue of the measurement of torsion as evidence of gravitational waves existence. With an experiment we gave some fundamentals studied in the gravitation theories, but with a modern mathematical study on invariants as are the twistors and spinors used to microscopic and microscopic field theory.

However, the limitations of our purely electronic devices only let see and interpret using the arguments of geometry, certain traces of electronic signals of the torsion evidence considering an electromagnetic field determined in certain voltage range and a movement of cylindrical trajectory, which as we know, is the constant torsion. However, this verifies Conjecture 2.1 and Theorem 3.1 established in other studies in theoretical physics and mathematical physics. Likewise, the methods and results of the research are on parallel themes and related to the gravity (no gravity precisely), considering this method as analogous to detect gravity waves but in this case to detect waves of torsion in an indirect way.

## Appendix. (A) The experimental data table to the cycles of magnetic dilaton displaced along the cylindrical spiral movement

| Degrees | Gs |
|---|---|
| 15 | 30.49 |
| 14 | 30.62 |
| 13 | 30.75 |
| 12 | 30.87 |
| 11 | 30.98 |
| 10 | 31.08 |
| 9 | 31.17 |
| 8 | 31.25 |
| 7 | 31.33 |
| 6 | 31.39 |
| 5 | 31.44 |
| 4 | 31.49 |
| 3 | 31.52 |
| 2 | 31.54 |
| 1 | 31.56 |
| 0 | 31.57 |
| 259 | 31.56 |
| 258 | 31.54 |
| 257 | 31.52 |
| 256 | 31.49 |
| 255 | 31.44 |
| 254 | 31.39 |
| 253 | 31.33 |
| 252 | 31.25 |
| 251 | 31.17 |
| 250 | 31.08 |
| 249 | 30.98 |
| 248 | 30.87 |
| 247 | 30.75 |
| 246 | 30.62 |
| 245 | 30.49 |

## Appendix. (B) Voltage that corresponds to proximity between magnetic dilaton (magnetic sphere) and sensor

| Voltage | Frequency |
|---|---|
| 0.01 | 26 |
| 0.02 | 122 |

| Voltage | Frequency |
|---|---|
| 0.03 | 155 |
| 0.04 | 165 |
| 0.05 | 174 |
| 0.07 | 205 |
| 0.08 | 220 |
| 0.09 | 252 |
| 0.10 | 275 |
| 0.11 | 303 |
| 0.12 | 324 |
| 0.13 | 338 |
| 0.14 | 344 |
| 0.15 | 365 |
| 0.16 | 380 |
| 0.17 | 404 |
| 0.18 | 422 |
| 0.19 | 443 |
| 0.20 | 457 |
| 0.21 | 489 |
| 0.22 | 495 |
| 0.23 | 502 |
| 0.24 | 530 |
| 0.25 | 542 |
| 0.26 | 559 |
| 0.25 | 548 |
| 0.24 | 530 |
| 0.23 | 503 |
| 0.22 | 495 |
| 0.21 | 483 |
| 0.20 | 457 |
| 0.19 | 443 |
| 0.18 | 422 |
| 0.17 | 404 |
| 0.16 | 380 |
| 0.15 | 265 |
| 0.14 | 344 |
| 0.13 | 338 |
| 0.12 | 324 |
| 0.11 | 303 |
| 0.10 | 275 |
| 0.09 | 252 |

| Voltage | Frequency |
|---|---|
| 0.08 | 220 |
| 0.07 | 205 |
| 0.05 | 174 |
| 0.04 | 165 |
| 0.03 | 155 |
| 0.02 | 122 |
| `0.01 | 96 |

## Author details

Francisco Bulnes[1]*, Isaías Martínez[2], Omar Zamudio[2] and Edgar Navarro[2]

1 IINAMEI, Research Department in Mathematics and Engineering, TESCHA, Mexico

2 IINAMEI A.C. (Investigación Internacional Avanzada en Matemáticas e Ingeniería), Mexico

*Address all correspondence to: gruposlie@yahoo.com.mx

## References

[1] Bulnes F. Detection and measurement of quantum gravity by a curvature energy sensor: H-states of curvature energy. In: Uzunov D, editor. Recent Studies in Perturbation Theory. Rijeka, Croatia: IntechOpen; 2017. Available from: https://www.intechopen.com/books/recent-studies-in-perturbation-theory/detection-and-measurement-of-quantum-gravity-by-a-curvature-energy-sensor-h-states-of-curvature-ener; https://doi.org/10.5772/68026

[2] Bulnes F, Stropovsvky Y, Rabinovich I. Curvature Energy and Their Spectrum in the Spinor-Twistor Framework: Torsion as Indicium of Gravitational Waves. Journal of Modern Physics. 2017;**8**:1723-1736. DOI: 10.4236/jmp.2017.810101

[3] Bulnes F. Electromagnetic Gauges and Maxwell Lagrangians applied to the determination of curvature in the space-time and their applications. Journal of Electromagnetic Analysis and Applications. 2012;**4**(6):252-266. DOI: 10.4236/jemaa.2012.46035

[4] Bulnes F. Gravity, curvature and energy: Gravitational field intentionality to the cohesion and union of the universe. In: Zouaghi T, editor. Gravity—Geoscience Applications, Industrial Technology and Quantum Aspect. London, UK: IntechOpen; 20 December 2017. DOI: 10.5772/intechopen.71037. Available from: https://www.intechopen.com/books/gravity-geoscience-applications-industrial-technology-and-quantum-aspect/gravity-curvature-and-energy-gravitational-field-intentionality-to-the-cohesion-and-union-of-the-uni

[5] Bulnes F, Martínez I, Mendoza A, Landa M. Design and development of an electronic sensor to detect and measure curvature of spaces using curvature energy. Journal of Sensor Technology. 2012;**2**(3):116-126. DOI: 10.4236/jst.2012.23017

[6] Bulnes F, Martínez I, Zamudio O, Negrete G. Electronic sensor prototype to detect and measure curvature through their curvature energy. Science Journal of Circuits, Systems and Signal Processing. 2015;**4**(5):41-54. DOI: 10.11648/j.cssp.20150405.12

[7] Bulnes F, Martínez I, Zamudio O. Fine curvature measurements through curvature energy and their gauging and sensoring in the space. In: Yurish SY, editor. Spain: Advances in Sensors Reviews 4, IFSA; 2016

## Chapter 3

# Brans-Dicke Solutions of Stationary, Axially Symmetric Spacetimes

*Pınar Kirezli Uludağ*


## Abstract

One of the most known alternative gravitational theories is Brans-Dicke (BD) theory. The theory offers a new approach by taking a scalar field $\phi$ instead of Newton's gravitational constant G. Solutions of the theory are under consideration and results are discussed in many papers. Stationary, axially symmetric solutions become important because gravitational field of celestial objects can be described by such solutions. Since obtaining exact solutions of BD is not an easy task, some solution-generating techniques are proposed. In this context, some solutions of Einstein general relativity, such as black hole or wormhole solutions, are discussed in BD theory. Indeed, black hole solutions in BD theory are not fully understood yet. Old and new such solutions and their analysis will be reviewed in this chapter.


**Keywords:** Brans-Dicke, stationary symmetric

## 1. Introduction

Einstein's theory of general relativity (GR), which is undoubtedly one of the greatest theories of the last century, is still being tried to be understood. Recently, the theory is supported by the observations of the gravitational waves which are observed by LIGO and Virgo collaboration [1]. On the other hand, GR may have some problems regarding defining gravity accurately at all scales. One of the problems that GR faced was that it could not fully describe the accelerated expansion of the universe [2–4] without unknown materials, i.e., dark matter and dark energy. Although, in order to understand the theory and satisfy the scientific cruosity, GR is modified with higher-order Ricci scalar [5, 6] soon after the theory is published, this modifications were not paid attention. The pioneer of studies on scalar-tensor theory were done by Brans and Dicke [7] by changing Newton's gravitational constant $G$ with a scalar field $\phi = 1/G$. In order to understand the BD theory, several experimental tests of GR are studied, and they are summarized in [8]. Additionally, it has been shown that BD theory can satisfy accelerated expansion of the universe with small and negative values of BD parameter $\omega$ [9, 10]. But these values of BD parameter cannot satisfy the solutions of our solar system and latest CMB datas. Extended BD theories which include a potential for the BD field are allowed to construct a number of analytic approximations [11]. Although, in the beginning, we construct our solutions with a potential, then we choose zero potential which is usually called as massless BD theory, in order to make it more simple.

Obtaining exact solutions of any theory is important in order to make comparison with observations or in order to obtain the results of the theory under consideration. Stationary, axially symmetric solutions are one of the important classes of these solutions, since the gravitational field of compact celestial objects such as stars, galaxies, and black holes can be represented by such solutions. Due to the complexity of the field equations, some solution-generating techniques are constructed. Obtaining Ernst BD equations is one of the most known of these techniques [12–14]. Also, Nayak and Tiwari [15] obtained vacuum stationary, axially symmetric BD solutions and generalized Maxwell field by Rai and Singh [16]. Their theory depends on finding out the relation between the field equations of BD and GR theories. After defining this relation, the corresponding BD solution of any known GR solution can be obtained. This method, which we call the Tiwari-Nayak-Singh-Rai (TNSR) method, is the most direct one. Instead of the one-parameter solution, which is called TNSR method, solutions with two parameters were constructed in [17]. A brief summary of [17] will be provided in the rest of the chapter.

The outline of the chapter will be as follow; in Section 2, we review the BD field equations and explain the Ernst equations and extended TNSR method. In Section 3, we study several solutions in order to understand how the extended TNSR method works. In addition, we mention the GR limit of the BD solutions.

## 2. Field equations of Brans-Dicke theory

In general a four-dimensional Brans-Dicke action with matter in Jordan frame is given by

$$S = \int d^4x\sqrt{-g}\left(\phi R - \frac{\omega(\phi)}{\phi} g^{\mu\nu}\partial_\mu\phi\partial_\nu\phi - V(\phi)\right) + S_M \tag{1}$$

where $8\pi G = c = 1$, $\phi$ is scalar field, $\omega(\phi)$ is BD coupling, $R$ is Ricci scalar, $V(\phi)$ is a potential BD field, and $S_M$ is action of the matter.

In many cases and origin of the BD theory, BD coupling $\omega(\phi)$ is choosen a constant as $\omega$. On the other hand, for different coupling function has a different modified gravity theory. One gets variations with respect to metric $g_{\mu\nu}$ and scalar field $\phi$

$$\begin{aligned} R_{\mu\nu} - \frac{1}{2}g_{\mu\nu}R = & \frac{T_{\mu\nu}}{\phi} - \frac{\omega(\phi)}{\phi^2}\left(\phi_{,\mu}\phi_{,\nu} - \frac{1}{2}g_{\mu\nu}\phi^{,\lambda}\phi_{,\lambda}\right) \\ & - \frac{1}{\phi}\left(\phi_{;\mu;\nu} - g_{\mu\nu}\Box\phi\right) - g_{\mu\nu}\frac{V(\phi)}{\phi} \end{aligned} \tag{2}$$

$$(2\omega + 3)\Box\phi = T - 2V + \phi\frac{dV}{d\phi} - \phi^{,\lambda}\phi_{,\lambda}\frac{d\omega(\phi)}{d\phi} \tag{3}$$

$T_{\mu\nu}$ represents energy-momentum tensor of the matter and $T$ is its trace and $\Box$ is d'Alembert operator with respect to full metric. The abstract index notation (i.e., $g_{\mu\nu}, T_{\mu\nu}$) is used in order to show component of what we concern with. Greek indices run over the spacetime manifold, starting with time component $t$ and space components $r, \theta, \varphi$ in this work. For example, if we want to figure out $t, t$ component of Einstein tensor ($G_{\mu\nu} = R_{\mu\nu} - \frac{1}{2}g_{\mu\nu}R$) from Eq. (2), we get

$$G_{tt} = \frac{T_{tt}}{\phi} - \frac{\omega(\phi)}{\phi^2}\left(\phi_{,t}\phi_{,t} - \frac{1}{2}g_{tt}\phi^{,\lambda}\phi_{,\lambda}\right) - \frac{1}{\phi}\left(\phi_{;t;t} - g_{tt}\Box\phi\right) - g_{tt}\frac{V(\phi)}{\phi} \tag{4}$$

where the repeated index $\lambda$ means summation, such as $\phi^{,\lambda}\phi_{,\lambda} = \phi^{,t}\phi_{,t} + \phi^{,\rho}\phi_{,\rho} + \phi^{,\theta}\phi_{,\theta} + \phi^{,\varphi}\phi_{,\varphi}$.

Conservation of the energy-momentum tensor of the matter leads to

$$T^{\mu}_{\nu;\nu} = 0. \tag{5}$$

Furthermore, we set BD coupling $\omega(\phi) \to \omega$ and scalar potential $V(\phi) = 0$ in the rest of the paper for simplicity and to easily obtain the field equations. Also, non-vanishing scalar potential and $\omega(\phi)$ are mostly used for cosmological solutions.

### 2.1 BD solution with electromagnetic field

A four-dimensional general stationary, axially symmetric spacetime can be represented with a metric in cylindrical coordinates in the canonical form as

$$ds^2 = -e^{2U}\left(dt^2 + \mathcal{A}d\varphi\right)^2 + e^{2(K-U)}\left(d\rho^2 + dz^2\right) + e^{-2U}W^2 d\varphi^2 \tag{6}$$

where all the metric functions depend on the coordinates $\rho$ and $z$. We shall consider the field content described by Maxwell field such that the energy-momentum tensor is

$$T_{\mu\nu} = 2\left(F_{\mu}{}^{\alpha}F_{\nu\alpha} - \frac{1}{4}g_{\mu\nu}F^{\alpha\beta}F_{\alpha\beta}\right) \tag{7}$$

and we study on the potential one form which shares the symmetry of the metric (6) as

$$A = A_0(\rho, z) + A_3(\rho, z) \tag{8}$$

with Maxwell equation

$$F^{\mu\nu}{}_{;\nu} = 0. \tag{9}$$

For simplicity we define BD field Eq. (2) as

$$Gdn_{\mu\nu} = G_{\mu\nu} - \frac{\omega}{\phi^2}\left(\phi,_{\mu}\phi,_{\nu} - \frac{1}{2}g_{\mu\nu}\phi^{,\alpha}\phi_{,\alpha}\right) - \frac{1}{\phi}\left(\phi_{;\mu;\nu} - g_{\mu\nu}\Box\phi\right) - \frac{2T_{\mu\nu}}{\phi} = 0 \tag{10}$$

and some of the field equations of the metric (6) become

$$Gdn^{\rho}_{\rho} + Gdn^{z}_{z} = \nabla^2(\phi W) = 0 \tag{11}$$

$$Gdn^{t}_{\varphi} = 4\left(\mathcal{A}\left(\vec{\nabla}A_0\right)^2 - \vec{\nabla}A_0\vec{\nabla}A_3\right) + We^{-2U}\vec{\nabla}.\left(\frac{e^{4U}\phi\vec{\nabla}\mathcal{A}}{W}\right) = 0 \tag{12}$$

$$Gdn_t^t + Gdn_\varphi^\varphi = 2\omega W^2\left(\vec{\nabla}\phi\right)^2 + 2\phi W\vec{\nabla}\left(W\vec{\nabla}\phi\right) - \phi^2 e^{4U}\left(\vec{\nabla}\mathcal{A}\right)^2 + 4W^2\phi^2\nabla^2 K$$
$$-4\phi^2 W^2\vec{\nabla}\left(W\vec{\nabla}U\right) + 2\phi^2 W\left[W\left(\vec{\nabla}U\right)^2 + \nabla^2 W\right] = 0 \qquad (13)$$

$$Gdn_t^t - Gdn_\varphi^\varphi = e^{-2U}W^2\left(\vec{\nabla}A_0\right)^2 - e^{2U}\left[\mathcal{A}^2\left(\vec{\nabla}A_0\right)^2 - \left(\vec{\nabla}A_3\right)^2\right]$$
$$-\mathcal{A}W\vec{\nabla}.\left(\frac{e^{4U}\phi\vec{\nabla}\mathcal{A}}{W}\right) - e^{4U}\phi\left(\vec{\nabla}\mathcal{A}\right)^2 + W\left[2\vec{\nabla}.\left(\phi W\vec{\nabla}U\right) - \vec{\nabla}.\left(\phi\vec{\nabla}W\right)\right] = 0$$
$$(14)$$

where $\vec{\nabla} f = \partial_\rho f, \partial_z f$. The field equation of (3) becomes

$$(2\omega + 3)\vec{\nabla}.\left(W\vec{\nabla}\phi\right) = 0. \qquad (15)$$

Using this result and Eq. (11) we obtain the last term in Eq. (14) vanishes except $\omega = -\frac{3}{2}$. Maxwell equation of (9) becomes

$$ME^t = \vec{\nabla}.\left[e^{-2U}W\vec{\nabla}A_0 + \frac{\mathcal{A}e^{2U}}{W}\left(\vec{\nabla}A_3 - \mathcal{A}\vec{\nabla}A_0\right)\right] = 0 \qquad (16)$$

$$ME^\varphi = \vec{\nabla}.\left[\frac{e^{2U}}{W}\left(\mathcal{A}\vec{\nabla}A_0 - \vec{\nabla}A_3\right)\right] = 0 \qquad (17)$$

BD field and Maxwell equations for stationary, axially spacetime are more complicated to obtain exact or approximate solutions. Even for GR, some solution-generating techniques are used because of this complexity. Firstly, we introduce Ernst equations obtained from BD field equations.

## 2.2 Ernst equations

The original Ernst equations in GR with the presence of Maxwell field are;

$$\left(\varepsilon + \varepsilon^* + |\Phi|^2\right)\nabla^2\varepsilon = 2\left(\vec{\nabla}\varepsilon + 2\Phi^*\vec{\nabla}\Phi\right)\vec{\nabla}\varepsilon \qquad (18)$$

$$\left(\varepsilon + \varepsilon^* + |\Phi|^2\right)\nabla^2\Phi = 2\left(\vec{\nabla}\varepsilon + 2\Phi^*\vec{\nabla}\Phi\right)\vec{\nabla}\Phi \qquad (19)$$

where $\varepsilon$ and $\Phi$ redefine potentials. Some exact solutions are obtained, integrating these equations. Our aim is to obtain BD-Maxwell field equations in this form. First, we rewrite the metric Ansatz:

$$ds^2 = -\alpha e^{\Omega/2}(dt + \mathcal{A}d\varphi)^2 + \alpha e^{-\Omega/2}d\varphi^2 + \frac{e^{2\nu}}{\sqrt{\alpha}}\left(d\rho^2 + dz^2\right) \qquad (20)$$

which simplifies the forthcoming equations considerably. Metric (6) and metric (20) relations are given:

$$W = \alpha, \qquad U = \frac{\Omega}{4} + \frac{1}{2}\ln\alpha, \qquad K = \frac{\Omega}{4} + \frac{1}{4}\ln\alpha + \nu. \qquad (21)$$

Defining an operator as $\vec{\nabla} f = \partial_\rho f, \partial_z f$, BD-Maxwell field equations become

$$EE_\rho^\rho + EE_z^z = \nabla^2(\phi\alpha) = 0 \tag{22}$$

$$EE_t^\varphi = \frac{1}{2}\vec{\nabla}.\left(\alpha\phi e^\Omega \vec{\nabla}\mathcal{A}\right) - \frac{2e^{\Omega/2}}{\mu_0}\left[\mathcal{A}\left(\vec{\nabla}A_0\right)^2 - \vec{\nabla}A_0.\vec{\nabla}A_3\right] = 0 \tag{23}$$

$$\begin{aligned} EE_t^t + EE_\varphi^\varphi = \frac{1}{8} \quad & \alpha\phi\left(\vec{\nabla}\Omega\right)^2 + 2\alpha\phi\nabla^2\nu - \frac{1}{2}\alpha\phi e^\Omega\left(\vec{\nabla}\mathcal{A}\right)^2 \\ & \frac{1}{2}\nabla^2(\alpha\phi) + \alpha\left[\frac{3}{2}\nabla^2\phi + \frac{m}{\phi}\left(\vec{\nabla}\phi\right)^2\right] = 0 \end{aligned} \tag{24}$$

$$\begin{aligned} EE_t^t - EE_\varphi^\varphi = \frac{2}{\mu_0}&\left[e^{\Omega/2}\left(\left(\vec{\nabla}A_0\right)^2 - \left(\vec{\nabla}A_3\right)^2\right) - e^{-\Omega/2}\left(\vec{\nabla}A_0\right)^2\right] \\ \frac{1}{2}\vec{\nabla}.\left(\alpha\phi\vec{\nabla}\Omega\right) &+ e^\Omega\alpha\phi\left(\vec{\nabla}\mathcal{A}\right)^2 + \mathcal{A}\vec{\nabla}.\left(e^\Omega\alpha\phi\vec{\nabla}\mathcal{A}\right) = 0 \end{aligned} \tag{25}$$

$$(3+2\omega)\vec{\nabla}.\left(\alpha\vec{\nabla}\phi\right) = 0 \tag{26}$$

From the equation of (9), we can write Maxwell equations as

$$E^t = \vec{\nabla}\left[e^{-\Omega/2}\vec{\nabla}A_0 + \mathcal{A}e^{\Omega/2}\left(\vec{\nabla}A_3 - \mathcal{A}\vec{\nabla}A_0\right)\right] = 0 \tag{27}$$

$$E^\varphi = \vec{\nabla}\left[e^{\Omega/2}\left(\vec{\nabla}A_3 - \mathcal{A}\vec{\nabla}A_0\right)\right] = 0 \tag{28}$$

With a new potential from the last Eq. (28),

$$\vec{e}_\varphi \times \vec{\nabla}\tilde{A}_3 = e^{\Omega/2}\left(\vec{\nabla}A_3 - \mathcal{A}\vec{\nabla}A_0\right),$$

$$\vec{e}_\varphi \times \vec{e}_\varphi \times \vec{\nabla}\tilde{A}_3 = e^{\Omega/2}\left(\vec{e}_\varphi \times \vec{\nabla}A_3 - \mathcal{A}\vec{e}_\varphi \times \vec{\nabla}A_0\right),$$

$$\vec{\nabla}.\left[\vec{e}_\varphi \times \vec{\nabla}A_3\right] = \vec{\nabla}.\left[\mathcal{A}\vec{e}_\varphi \times \vec{\nabla}A_0 - e^{-\Omega/2}\vec{\nabla}\tilde{A}_3\right] = 0 \tag{29}$$

With the new potential, the other Maxwell Eq. (27) is written:

$$\vec{\nabla}\left[e^{-\Omega/2}\vec{\nabla}A_0 + \mathcal{A}\vec{e}_\varphi \times \vec{\nabla}\tilde{A}_3\right] = 0. \tag{30}$$

We define a new complex potential

$$\Phi = A_0 + i\tilde{A}_3 \tag{31}$$

and Maxwell equations become

$$\vec{\nabla}.\left[e^{-\Omega/2}\vec{\nabla}\Phi - i\mathcal{A}\vec{e}_\varphi \times \vec{\nabla}\Phi\right] = 0 \tag{32}$$

where the real part is equal to Eq. (27) and the complex part is equal to (28). Eq. (23) is written in the form

$$\vec{\nabla}.\left[e^\Omega\phi\alpha\vec{\nabla}\mathcal{A} - 2\vec{e}_\varphi \times Im\left(\Phi^*\vec{\nabla}\Phi\right)\right] = 0 \tag{33}$$

where $1/\mu_0 = 1$. From the last equation, we can define a new potential like

$$\begin{aligned}\vec{e}_\varphi \times \vec{\nabla} h &= e^{\Omega}\phi\alpha\vec{\nabla}\mathcal{A} - 2\vec{e}_\varphi \times Im\left(\Phi^* \vec{\nabla}\Phi\right)\\ \vec{e}_\varphi \times \vec{e}_\varphi \times \vec{\nabla} h &= e^{\Omega}\phi\alpha\vec{e}_\varphi \times \vec{\nabla}\mathcal{A} - 2\vec{e}_\varphi \times \vec{e}_\varphi \times Im\left(\Phi^* \vec{\nabla}\Phi\right)\\ \vec{e}_\varphi \times \vec{\nabla}\mathcal{A} &= \frac{-e^{-\Omega}}{\phi\alpha}\left(\vec{\nabla} h + 2Im\left(\Phi^* \vec{\nabla}\Phi\right)\right) \end{aligned} \tag{34}$$

and the equation of (23) becomes

$$\vec{\nabla}.\left[\frac{e^{-\Omega}}{\phi\alpha}\left(\vec{\nabla} h + 2Im\left(\Phi^* \vec{\nabla}\Phi\right)\right)\right] = 0. \tag{35}$$

A new function is defined

$$f = e^{\Omega/2}\alpha\phi \tag{36}$$

and (25) is obtained:

$$\frac{f}{\alpha\phi}\vec{\nabla}.\left(\alpha\phi\vec{\nabla} f\right) - \vec{\nabla} f.\vec{\nabla} f - \frac{f^2}{\alpha\phi}\nabla^2(\alpha\phi) = 2f\vec{\nabla}\Phi.\vec{\nabla}\Phi^* - \left[\vec{\nabla} h + 2Im\left(\Phi^* \vec{\nabla}\Phi\right)\right]^2 \tag{37}$$

If one introduces the complex function

$$\varepsilon = f - |\Phi|^2 + ih \tag{38}$$

field equations of (23) and (25) and Maxwell Eqs. (27) and (28) can be writen as

$$\left(\varepsilon + \varepsilon^* + |\Phi|^2\right)\frac{1}{\alpha\phi}\vec{\nabla}.\left(\alpha\phi\vec{\nabla}\varepsilon\right) = \left(\vec{\nabla}\varepsilon + 2\Phi^*\vec{\nabla}\Phi\right).\vec{\nabla}\varepsilon + Re^2\left(\varepsilon + |\Phi|^2\right)\frac{\nabla^2(\alpha\phi)}{\alpha\phi} \tag{39}$$

$$\left(\varepsilon + \varepsilon^* + |\Phi|^2\right)\frac{1}{\alpha\phi}\vec{\nabla}.\left(\alpha\phi\vec{\nabla}\Phi\right) = \left(\vec{\nabla}\varepsilon + 2\Phi^*\vec{\nabla}\Phi\right).\vec{\nabla}\Phi. \tag{40}$$

The last term of Eq. (39) becomes zero from the field equation of (22). Additionally, this field equation permits us to choose $\alpha\phi = \rho$ which reduces the field equations to Ernst equations of (18) and (19). Besides, BD-Maxwell Ernst equations do not include the field equations of (24) and (26). For obtaining solutions, the chosen appropriate physical potentials for $\varepsilon$ and $\Phi$ which satisfy the Ernst equations are not sufficent. They must also satisfy the field equations of (24) and (26). For that reason, we introduce another method for solution of BD-Maxwell equations of stationary, axially symmetric spacetimes.

### 2.3 Extended Tiwari-Nayak-Rai-Singh method

In this subsection, we try to analyze how to obtain BD-Maxwell solution from a known Einstien-Maxwell solution for stationary, axially symmetric spacetime. We start with writing a metric as

$$ds^2 = -e^{2U_e}(dt + \mathcal{A}_e d\varphi)^2 + e^{2K_e - 2U_e}\left(d\rho^2 + dz^2\right) + W_e^2 e^{-2U_e} d\varphi^2 \tag{41}$$

where the subscript refers to Einstein metric functions. The first of the field equations in GR is obtained:

$$G^\rho_\rho + G^z_z = \nabla^2 W_e = 0, \tag{42}$$

$$Gd^\phi_t = W_e \vec{\nabla}\left(\frac{e^{4U_e}\vec{\nabla}\mathcal{A}_e}{W_e}\right) + 2e^{2U_e}\left(\vec{\nabla}A_0\vec{\nabla}A_3 - \mathcal{A}_e\left(\vec{\nabla}A_0\right)^2\right). \tag{43}$$

When we choose the Einstein and BD field share the same Maxwell field which means $A_0$ and $A_3$ are the same for GR and BD, Eqs. (11) and (12) become more similar to Eqs. (42) and (43). From the first equations, we can write $W_e = \phi W$, and it permit us to choose $W = W_e^k$ and $\phi = W^{1-k}$. In the next step, we take the metric function $\mathcal{A}$ of BD and $\mathcal{A}_e$ of GR are the same which satisfy the relation between $U = U_e - \frac{1}{2}\ln\phi$. Finally, by using the equation $Gdn^t_t + Gdn^\varphi_\varphi - \left(Gd^t_t + Gd^\varphi_\varphi\right) = 0$, we obtain $K = K_e + \frac{1}{4}(2\omega - 1 - k(2\omega+3))\ln\phi$. Previous studies took the metric functions $K_e = K$ where $k = \frac{2\omega-1}{2\omega+3}$ [15]. We can summarize that, if there is a GR solution in the form of metric (6), it has a corresponding BD solution with transformation of the metric function as

$$\mathcal{A} = \mathcal{A}_e, \qquad W = W_e^k, \qquad \phi = W_e^{1-k}, \tag{44}$$

$$U = U_e - \frac{1}{2}\ln\phi, \tag{45}$$

$$K = K_e + \frac{1}{4}(2\omega - 1 - k(2\omega+3))\ln\phi. \tag{46}$$

Additionally, if the seed solution is in the form of Eq. (20), the corresponding BD solution may be obtained by the transformation as

$$\alpha = \alpha_e, \qquad \phi = \alpha_e^{1-k}, \qquad \Omega = \Omega_e \tag{47}$$

$$\mathcal{A} = \mathcal{A}_e \qquad \nu = \nu_e + \frac{2\omega - k(2\omega+3)}{4}\ln\phi. \tag{48}$$

## 3. Examples of BD solutions with extended TNSR method

### 3.1 BD solution of Kerr-Taub-NUT metric

We also know Kerr-Taub-NUT (KTN) vacuum solution is

$$\begin{aligned} ds^2 &= -\frac{1}{\rho^2}(\Delta - a^2\sin^2\theta)dt^2 + \frac{2}{\rho^2}\left[\Delta\alpha - a(\rho^2 + a\alpha)\sin^2\theta\right]dtd\phi \\ &\quad + \frac{1}{\rho^2}\left[\left(\rho^2 + a\alpha\right)^2\sin^2\theta\right]d\phi^2 + \rho^2\left[\frac{dr^2}{\Delta} + d\theta^2\right] \end{aligned} \tag{49}$$

where

$$\begin{aligned} \Delta &= r^2 - 2Mr + a^2 - n^2, \qquad \rho^2 = r^2 + (n + a\cos\theta)^2 \\ \alpha &= a\sin^2\theta - 2n\cos\theta \end{aligned}$$

and $n$ is NUT parameter, $a$ is rotational parameter, and $M$ is the mass. By doing a coordinate transformation like $r = e^R + M + \left(\frac{M^2 - a^2 + n^2}{4}\right)e^{-R}$, Kerr-Taub-NUT metric becomes

$$\begin{aligned} ds^2 &= -\frac{1}{\rho^2}(\Delta - a^2 \sin^2\theta)dt^2 + \frac{2}{\rho^2}\left[\Delta\alpha - a\left(\rho^2 + a\alpha\right)\sin^2\theta\right]dtd\phi \\ &+ \frac{1}{\rho^2}\left[\left(\rho^2 + a\alpha\right)^2 \sin^2\theta\right]d\phi^2 + \rho^2\left[dR^2 + d\theta^2\right]. \end{aligned} \tag{50}$$

which is similar to metric (6) and the functions are

$$\Delta = L^2 - 2ML + a^2 - n^2, \qquad \rho^2 = L^2 + (n + a\cos\theta)^2$$

where $L = e^R + M + \left(\frac{M^2 - a^2 + n^2}{4}\right)e^{-R}$. From this metric we can easily find that

$$\begin{aligned} e^{2U_e} &= \frac{1}{\rho^2}\left(\Delta - a^2 \sin^2\theta\right) \\ \mathcal{A}_e &= \frac{a\left(\rho^2 + a\alpha\right)^2 \sin^2\theta - \Delta\alpha}{\Delta - a^2\sin^2\theta} \\ e^{2K_e} &= \Delta - a^2\sin^2\theta \\ W_e &= \sqrt{\Delta}\sin\theta \end{aligned}$$

From Eqs. (44) to (46) and by doing a coordinate transformation again, the solution looks

$$\begin{aligned} ds^2 = \left[\sqrt{\Delta}\sin\theta\right]^{k-1}\Bigg[&-\frac{\Delta - a^2\sin^2\theta}{\rho^2}dt^2 - \frac{2}{\rho^2}\left[a\left(\rho^2 + a\alpha\right)\sin^2\theta - \Delta\alpha\right]dtd\phi \\ &+\left(\left(\rho^2 + a\alpha\right)^2\sin^2\theta - \Delta\alpha^2\right)d\phi^2\Bigg] + \left[\sqrt{\Delta}\sin\theta\right]^{\frac{(1-k)}{2}(1+2\omega-k(2\omega+3))}\left(\frac{\rho^2 dr^2}{\Delta} + \rho^2 d\theta^2\right) \end{aligned} \tag{51}$$

### 3.2 BD solution of Kerr-Newman-Taub-NUT metric

The solution for Kerr-Newman-Taub-NUT (KNTN) metric is the same as Eq. (49), but the functions are

$$\begin{aligned} \Delta &= r^2 - 2Mr + a^2 - n^2 + Q^2, \qquad \rho^2 = r^2 + (n + a\cos\theta)^2 \\ \alpha &= a\sin^2\theta - 2n\cos\theta \end{aligned}$$

which correspond to mass of KTN solution which is $M = M - Q^2/2r$. By doing the same procedure for KTN solution, we can easily obtain KNTN solution, but there are some differences such as for metric transformation, $r = e^R + M + \left(\frac{M^2 - a^2 + n^2 - Q^2}{4}\right)e^{-R}$ is used. Brans-Dicke solution of KNTN becomes

$$\begin{aligned} ds^2 = \left[\sqrt{\Delta}\sin\theta\right]^{k-1}\Bigg[&-\frac{\Delta - a^2\sin^2\theta}{\rho^2}dt^2 - \frac{2}{\rho^2}\left[a\left(\rho^2 + a\alpha\right)\sin^2\theta - \Delta\alpha\right]dtd\phi \\ &+\left(\left(\rho^2 + a\alpha\right)^2\sin^2\theta - \Delta\alpha^2\right)d\phi^2\Bigg] + \left[\sqrt{\Delta}\sin\theta\right]^{\frac{(1-k)}{2}(1+2\omega-k(2\omega+3))}\left(\frac{\rho^2 dr^2}{\Delta} + \rho^2 d\theta^2\right). \end{aligned} \tag{52}$$

This solution was obtained by a sigma-model theory in [18] where the parameters have a relation like $\alpha = (1-k)(2\omega+3)/4$. Moreover, analyses of the charged particle geodesics around this spacetime were discussed in [17] analytically.

### 3.3 BD solution of magnetized Kerr-Newman solution

Magnetized Kerr-Newman solution was found by Gibbons and his friends [19] as

$$ds^2 = H\left[-fdt^2 + \Sigma\left(\frac{dr^2}{\Delta} + d\theta^2\right)\right] + \frac{S\sin^2\theta}{H\Sigma}(d\varphi - \beta dt)^2 \tag{53}$$

$$A = \Phi_0 dt + \Phi_3(d\varphi - \beta dt). \tag{54}$$

where

$$\begin{aligned} &\Sigma = r^2 + a^2\cos^2\theta \qquad \Delta = r^2 + a^2 - 2mr + p^2 + q^2 \\ &f = \frac{\Delta\Sigma}{S} \qquad S = \left(r^2 + a^2\right)^2 - a^2\Delta\sin^2\theta \\ &\beta = \frac{a(2mr - p^2 - q^2)}{S} \end{aligned} \tag{55}$$

where $a$ is rotational parameter, $p$ is magnetic charge, and $q$ is electric charge of rotating black hole. We can easily compare (53) and (6) by doing $r = e^R + M + \frac{M^2-a^2-p^2-q^2}{4}e^{-R}$ transformation. GR metric functions become

$$e^{2U_e} = \frac{H^2 f\Sigma - S\beta^2\sin^2\theta}{H\Sigma} \qquad \mathcal{A}_e = \frac{S\beta\sin^2\theta}{H^2 f\Sigma - S\beta^2\sin^2\theta} \tag{56}$$

$$W_e = \sqrt{\Delta}\sin\theta \qquad e^{2K_e} = H^2 f\Sigma - S\beta^2\sin^2\theta. \tag{57}$$

where

$$\begin{aligned} &\Sigma = L^2 + a^2\cos^2\theta \qquad \Delta = L^2 + a^2 - 2mL + p^2 + q^2 \\ &f = \frac{\Delta\Sigma}{S} \qquad S = \left(L^2 + a^2\right)^2 - a^2\Delta\sin^2\theta \\ &\beta = \frac{a(2mL - p^2 - q^2)}{S} \end{aligned} \tag{58}$$

where $L = e^R + M + \frac{M^2-a^2-p^2-q^2}{4}e^{-R}$. Metric functions of BD solution of magnetized KN are obtained from Eqs. (44) to (46), and by inverse transformation of $R \to r$, metric of BD solution of magnetized KN becomes

$$\begin{aligned} ds^2 = &\left(\sqrt{\Delta}\sin\theta\right)^{k-1}\left[\left(-Hf + \frac{S\beta^2\sin^2\theta}{H\Sigma}\right)dt^2 - 2\frac{S\beta\sin^2\theta}{H\Sigma}dtd\varphi + \frac{S\sin^2\theta}{H\Sigma}d\varphi^2\right] \\ &\left(\sqrt{\Delta}\sin\theta\right)^{\frac{(1-k)}{2}(1+2\omega-k(2\omega+3))}H\Sigma\left(\frac{dr^2}{\Delta} + d\theta^2\right). \end{aligned} \tag{59}$$

## 4. GR limit of the solutions

According to the common belief, since BD parameter $\omega \to \infty$, the BD solutions reduce to corresponding GR ones. Contrary to this belief, several counter examples were presented in the literature [20–24]. In our study, the GR limit of the BD solutions is out of complexity. When $k \to 1$, BD transformation equations of (44)–(46) reduce to seed GR metric functions for any finite $\omega$ since scalar field $\phi$ becomes constant. It is obvious from the given examples that, as $k \to 1$, BD metrics reduce the corresponding GR ones.

## 5. Conclusion

In this section, we have studied to obtain corresponding BD or BD-Maxwell solution from any known solution of the Einstein or the Einstein-Maxwell theory for stationary, axially symmetric spacetimes in Jordan frame. First we present that, although several field equations of BD are not included by Ernst equations, BD field equations can be written in the form of Ernst Eqs. BD solutions can be obtained by selecting the appropriate physical potentials or by integrating Ernst equations, but it should be remembered that the equations which are not included in the Ernst equations should be provided.

In order to obtain BD solutions, we have constructed two parameter solution-generating techniques. It was seen that, in previous works, it was studied with one parameter. From any given seed GR solution of Eqs. (6) or (20), the corresponding BD solution can be obtained by the two parameter solution-generating techniques. In order to show how this method works, we have constructed several known solutions and also some new solutions for BD theory. We have also discussed the GR limit of these solutions.

## Author details

Pınar Kirezli Uludağ
Department of Physics, Namık Kemal University, Tekirdağ, Turkey

*Address all correspondence to: pkirezli@nku.edu.tr

## References

[1] Abbott BP et al. Observation of gravitational waves from a binary black hole merger. Physical Review Letters. 2016;**116**:061102

[2] Perlmutter S et al. Measurements of Ω and Λ from 42 high redshift supernovae. The Astrophysical Journal. 1999;**517**:565-586

[3] Perlmutter S, Turner MS, White M. Constraining dark energy with SNe Ia and large scale structure. Physical Review Letters. 1999;**83**:670-673

[4] Riess AG et al. Type Ia supernova discoveries at $z > 1$ from the hubble space telescope: Evidence for past deceleration and constraints on dark energy evolution. The Astrophysical Journal. 2004;**607**:665-687

[5] Weyl H. A new extension of relativity theory. Annalen der Physik. 1919;**59**:101-103

[6] Eddeington AS. The Mathematical Theory of Relativity. Cambridge: Cambridge University Press; 1923

[7] Brans C, Dicke R. Mach's principle and a relativistic theory of gravitation. Physics Review. 1961;**124**:925-935

[8] McHugh MP. The Brans-Dicke theory and its experimental tests. Fundamental Theories of Physics. 2016; **183**:163

[9] Banerjee N, Pavon D. Cosmic acceleration without quintessence. Physical Review D. 2001;**63**:043504

[10] Sen AA, Sen S, Sethi S. Dissipative fluid in Brans-Dicke theory and late time acceleration. Physical Review D. 2001;**63**:107501

[11] Lima NA, Ferreira PG. On the phenomenology of extended Brans-Dicke gravity. JCAP. 2016;**1601**:010. DOI: 10.1088/1475-7516/2016/01/010 [arXiv:1506.07771 [astro-ph.CO]]

[12] Chatterjee S, Banerji S. Acta Physica Polonica B. 1980;**11**:493

[13] Tsuchida T, Watanabe K. Comment on the exterior solutions and their geometry in scalar-tensor theories of gravity. Progress in Theoretical Physics. 1999;**101**:73-90

[14] Astorino M. Embedding hairy black holes in a magnetic universe. Physical Review D. 2013;**87**:084029

[15] Nayak BK, Tiwari RN. Class of the Brans-Dicke Maxwell fields. Physical Review D. 1976;**14**:2502-2504

[16] Singh T, Rai LN. Stationary axially symmetric Brans-Dicke-Maxwell fields. General Relativity and Gravitation. 1979;**11**:37-43

[17] Kirezli P, Delice O. Stationary axially symmetric solutions in Brans-Dicke theory. Physics Review. 2015; **D92**:104045

[18] Park DH. Rotating black hole solutions in the Brans-Dicke theory. Journal of the Korean Physical Society. 2007;**51**:258-262

[19] Gibbons GW, Mujtaba AH, Pope CN. Ergoregions in magnetized black hole spacetimes. Classical and Quantum Gravity. 2013;**30**

[20] Baykal A, Delice O. Cylindrically symmetric Brans-Dicke-Maxwell solutions. General Relativity and Gravitation. 2009;**41**:267-285

[21] Romero C, Barros A. Does Brans-Dicke theory of gravity go over to the general relativity when omega − − − > infinity? Physics Letters A. 1993; **173**:243-246

[22] Banerjee N, Sen S. Does Brans-Dicke theory always yield general relativity in the infinite omega limit? Physical Review D. 1997;**56**:1334-1337

[23] Bhadra A, Nandi KK. Omega dependence of the scalar field in Brans-Dicke theory. Physical Review D. 2001; **64**:087501

[24] Faraoni V, Côté J. Two new approaches to the anomalous limit of Brans-Dicke theory to Einstein gravity. Physical Review D. 2019;**99**(6):064013

Section 3

# Tensors in Geometry and Continuum Media

Chapter 4

# Fluid Motion Equations in Tensor Form

*Dmitry Nikushchenko and Valery Pavlovsky*

## Abstract

In the current chapter, some applications of tensor analysis to fluid dynamics are presented. Governing equations of fluid motion and energy are obtained and analyzed. We shall discuss about continuity equation, equation of motion, and mechanical energy transport equation and four forms of energy equation. Finally, we shall talk about the divergence from transfer equations of different parameters of motion. The tensor form of equations has advantages over the component form: these are, first, compact writing of equations and, second, independency from reference frames, etc. Moreover, it allows to obtain new forms of equations on the basis of governing ones easily.

**Keywords:** stress tensor, Navier–stokes equation, energy, continuity, vorticity, divergence form

## 1. Introduction

The mathematical model of moving fluid includes a set of equations, which are usually written as transport equations of main physical parameters—density, velocity, energy, etc. These equations are conservation laws in fluid flows. Traditionally the component form of the equations is usually used, but at the same time, the componentless form (Gibbs approach) could be applied to obtain and transform these equations. In this chapter several main conservation laws are discussed and represented in tensor form, which has many advantages against usually used component form, like simplicity and compactness, independence on reference frames, less errors in transformations, etc. Below we obtain and analyze continuity and momentum equations and vorticity and energy transport equations, and we discuss also about the divergent form of transport Eqs.

## 2. Continuity equation

Continuity equation is the mass conservation law for a fluid flow and is presented as a scalar equation, which connects density $\rho$ and velocity of fluid particles $\vec{V}$, and for any liquid it could be written as

$$\frac{\partial \rho}{\partial t} + \left(\vec{V} \cdot \vec{\nabla}\right)\rho + \rho \vec{\nabla} \cdot \vec{V} = 0, \qquad (1)$$

where $\vec{\nabla} = \vec{e}_i \frac{\partial}{\partial x_i} = \vec{e}_1 \frac{\partial}{\partial x_1} + \vec{e}_2 \frac{\partial}{\partial x_2} + \vec{e}_2 \frac{\partial}{\partial x_3}$ is the Hamilton operator and the dot is a symbol of a scalar product.

Hereinafter, the Einstein summation convention is used by default.

It also can be written in two other equivalent forms [1, 2]:

$$\frac{d\rho}{dt} + \rho\vec{\nabla}\cdot\vec{V} = 0 \text{ and } \frac{\partial\rho}{\partial t} + \vec{\nabla}\cdot\rho\vec{V} = 0. \tag{2}$$

In the case of incompressible fluid, we could obtain its simplified form:

$$\vec{\nabla}\cdot\vec{V} = 0. \tag{3}$$

Let us apply gradient operator to continuity equation (Eq. (1)):

$$\vec{\nabla}\frac{\partial\rho}{\partial t} + \vec{\nabla}\left[\left(\vec{V}\cdot\vec{\nabla}\right)\rho\right] + \vec{\nabla}\left[\rho\vec{\nabla}\cdot\vec{V}\right] = 0.$$

As a result, we obtain vector equation:

$$\frac{\partial}{\partial t}\vec{\nabla}\rho + \vec{\nabla}\rho\cdot\vec{\nabla}\vec{V}^{T} + \vec{V}\cdot\vec{\nabla}\vec{\nabla}\rho + \left(\vec{\nabla}\cdot\vec{V}\right)\vec{\nabla}\rho + \rho\vec{\nabla}\left(\vec{\nabla}\cdot\vec{V}\right) = 0, \tag{4}$$

which could be written in a more compact form:

$$\frac{d}{dt}\vec{\nabla}\rho + \vec{\nabla}\vec{V}\cdot\vec{\nabla}\rho + \left(\vec{\nabla}\cdot\vec{V}\right)\vec{\nabla}\rho + \rho\vec{\nabla}\left(\vec{\nabla}\cdot\vec{V}\right) = 0 \tag{5}$$

or by a little bit different way:

$$\frac{d}{dt}\vec{\nabla}\rho + \vec{\nabla}\vec{V}\cdot\vec{\nabla}\rho + \vec{\nabla}\left(\rho\vec{\nabla}\cdot\vec{V}\right) = 0. \tag{6}$$

These equations contain gradient of vector $\vec{V}$ divergence, which [3] equal to

$$\vec{\nabla}\left(\vec{\nabla}\cdot\vec{V}\right) = \Delta\vec{V} + \vec{\nabla}\times\vec{\nabla}\times\vec{V}.$$

For incompressible fluid the left part of this relation is equal to zero; therefore for rotation of velocity vector, we can write:

$$\vec{\nabla}\times\vec{\nabla}\times\vec{V} = -\Delta\vec{V}. \tag{7}$$

In the case of compressible fluid in accordance with Eq. (6), we have additional terms in the right part of the equation:

$$\vec{\nabla}\times\vec{\nabla}\times\vec{V} = -\left(\Delta\vec{V} + \frac{1}{\rho}\frac{d}{dt}\vec{\nabla}\rho + \frac{1}{\rho}\vec{\nabla}\vec{V}\cdot\vec{\nabla}\rho + \frac{\left(\vec{\nabla}\cdot\vec{V}\right)}{\rho}\vec{\nabla}\rho\right). \tag{8}$$

Continuity equation can be also written in tensor form:

$$\frac{1}{3}tr\left(\frac{\partial\rho}{\partial t}\underline{E}\right) + tr\left(\vec{\nabla}\rho\vec{V}\right) = 0. \tag{9}$$

The tensor $\vec{\nabla}\rho\vec{V}$ can be represented as

$$\vec{\nabla}\rho\vec{V} = \vec{e}_k \frac{\partial}{\partial x_k}\rho V_j \vec{e}_j = \frac{\partial}{\partial x_k}\rho V_j \vec{e}_k \vec{e}_j = \\ = \left[\frac{\partial \rho}{\partial x_k} V_j + \rho \frac{\partial V_j}{\partial x_k}\right]\vec{e}_k \vec{e}_j = \vec{\nabla}\rho \otimes \vec{V} + \rho \vec{\nabla}\vec{V}. \tag{10}$$

Finally, continuity equation can be written in form

$$\frac{\partial \rho}{\partial t} + tr\left[\vec{\nabla}\rho \otimes \vec{V} + \rho\vec{\nabla}\vec{V}\right] = 0. \tag{11}$$

Convective derivatives of density and pressure (and any another scalar quantitatives) also can be written in tensor form:

$$\left(\vec{V}\cdot\vec{\nabla}\right)\rho = tr\left(\vec{V}\otimes\vec{\nabla}\rho\right); \quad \left(\vec{V}\cdot\vec{\nabla}\right)p = tr\left(\vec{V}\otimes\vec{\nabla}p\right),$$

i.e., convective derivative is equal to trace of corresponding tensor.

In addition, for the divergence of the product of scalar and vector functions, we can obtain the following relation:

$$\vec{\nabla}\cdot\rho\vec{V} = \left(\vec{\nabla}\rho\otimes\vec{V} + \rho\vec{\nabla}\vec{V}\right) : \underline{E}.$$

## 3. Equations of motion of fluid with constant and variable properties

The equation of a motion in terms of stress [4, 5] is

$$\rho\frac{d\vec{V}}{dt} = \vec{\nabla}\cdot\underline{\sigma} + \rho\vec{f}, \tag{12}$$

where $\vec{f}$ is the body force per unit mass and $\vec{\nabla}\cdot\underline{\sigma}$ is the divergence of stress tensor $\underline{\sigma}$. In accordance with Newton's law, tensor $\underline{\sigma}$ for an incompressible fluid is

$$\underline{\sigma} = -p\underline{E} + 2\mu\underline{S}, \tag{13}$$

where $p$ is the pressure; $\mu$ is the fluid shear (dynamic) viscosity; and $\underline{S} = \frac{1}{2}\left(\vec{\nabla}\vec{V} + \vec{\nabla}\vec{V}^T\right)$ is the rate of strain tensor. Due to relation $\vec{\nabla}\cdot\vec{V} = 0$, when $\mu \neq const$ divergence of stress tensor $\underline{\sigma}$ is written as

$$\vec{\nabla}\cdot\underline{\sigma} = -\vec{\nabla}p + \vec{\nabla}\mu\cdot 2\underline{S} + 2\mu\vec{\nabla}\cdot\underline{S} = -\vec{\nabla}p + \vec{\nabla}\mu\cdot 2\underline{S} + \mu\Delta\vec{V}. \tag{14}$$

Then equation of motion of incompressible fluid (Navier–Stokes equation) at $\mu \neq const$ is

$$\rho\frac{d\vec{V}}{dt} = -\vec{\nabla}p + \mu\Delta\vec{V} + \vec{\nabla}\mu\cdot 2\underline{S} + \rho\vec{f}. \tag{15}$$

Additional term $\vec{\nabla}\mu \cdot 2\underline{S}$ relates to changing of shear viscosity. In Cartesian coordinates Eq. (15) has the form:

$$\rho\frac{dV_i}{dt} = -\frac{\partial p}{\partial x_i} + \mu\Delta V_i + \frac{\partial \mu}{\partial x_j}\left(\frac{\partial V_j}{\partial x_i} + \frac{\partial V_i}{\partial x_j}\right) + \rho f_i.$$

In case of compressible fluid with variable viscosity, the equation will contain a term with divergence $\vec{\nabla} \cdot \vec{V}$, which is not equal to zero now. Rheological relation is this case has the form [2]:

$$\underline{\sigma} = -p\underline{E} - \frac{2}{3}\mu\left(\vec{\nabla} \cdot \vec{V}\right)\underline{E} + 2\mu\underline{S}. \tag{16}$$

Let us introduce the denotation:

$$p + \frac{2}{3}\mu\left(\vec{\nabla} \cdot \vec{V}\right) = p'; \tag{17}$$

then we can write Eq. (16) as

$$\underline{\sigma} = -p'\underline{E} + 2\mu\underline{S}. \tag{18}$$

Divergence of this tensor at $\mu \neq const$ is

$$\vec{\nabla} \cdot \underline{\sigma} = -\vec{\nabla}p' + \vec{\nabla}\mu \cdot 2\underline{S} + \mu\Delta\vec{V} + \mu\vec{\nabla}\left(\vec{\nabla} \cdot \vec{V}\right).$$

As a result, equation of motion of compressible fluid with variable viscosity has the form:

$$\rho\frac{d\vec{V}}{dt} = -\vec{\nabla}p' + \mu\Delta\vec{V} + \mu\vec{\nabla}\left(\vec{\nabla} \cdot \vec{V}\right) + \vec{\nabla}\mu \cdot 2\underline{S} + \rho\vec{f}, \tag{19}$$

in Cartesian coordinates

$$\rho\left[\frac{\partial V_i}{\partial t} + V_j\frac{\partial V_i}{\partial x_j}\right] = -\frac{\partial p'}{\partial x_i} + \mu\Delta V_i + \mu\frac{\partial}{\partial x_i}\left(\frac{\partial V_j}{\partial x_j}\right) + \frac{\partial \mu}{\partial x_j}\left(\frac{\partial V_j}{\partial x_i} + \frac{\partial V_i}{\partial x_j}\right) + \rho f_i.$$

If we represent fluid particle acceleration as the sum of local and convective terms, then (Eq. (19)) will take the form:

$$\rho\frac{\partial\vec{V}}{\partial t} + \rho\left(\vec{V} \cdot \vec{\nabla}\right)\vec{V} = -\vec{\nabla}p' + \mu\Delta\vec{V} + \mu\vec{\nabla}\left(\vec{\nabla} \cdot \vec{V}\right) + \vec{\nabla}\mu \cdot 2\underline{S} + \rho\vec{f}, \tag{20}$$

considering viscosity variability is especially important for turbulent flow modeling using the Boussinesq hypothesis with turbulent viscosity $\mu_t$.

Let us apply divergence operation to the Navier–Stokes equation for compressible fluid with variable viscosity. With this purpose we shall apply operation $\left(\vec{\nabla}\cdot\right)$ to each vector term of Eq. (20):

$$\vec{\nabla}\cdot\rho\frac{\partial\vec{V}}{\partial t}=\vec{\nabla}\rho\cdot\frac{\partial\vec{V}}{\partial t}+\rho\frac{\partial}{\partial t}\vec{\nabla}\cdot\vec{V};$$

$$\vec{\nabla}\cdot\rho\left(\vec{V}\cdot\vec{\nabla}\right)\vec{V}=\vec{\nabla}\rho\cdot\left[\left(\vec{V}\cdot\vec{\nabla}\right)\vec{V}\right]+\rho\vec{\nabla}\vec{V}:\vec{\nabla}\vec{V}+\rho\left(\vec{\nabla}\cdot\vec{V}\right)\left(\vec{V}\cdot\vec{\nabla}\right);$$

$$\vec{\nabla}\cdot\left(-\vec{\nabla}p'\right)=-\Delta p';$$

$$\vec{\nabla}\cdot\mu\Delta\vec{V}=\vec{\nabla}\mu\cdot\Delta\vec{V}+\mu\Delta\left(\vec{\nabla}\cdot\vec{V}\right);$$

$$\vec{\nabla}\cdot\mu\vec{\nabla}\left(\vec{\nabla}\cdot\vec{V}\right)=\vec{\nabla}\mu\cdot\vec{\nabla}\left(\vec{\nabla}\cdot\vec{V}\right)+\mu\Delta\left(\vec{\nabla}\cdot\vec{V}\right);$$

$$\vec{\nabla}\cdot\left(\vec{\nabla}\mu\cdot 2\underline{S}\right)=\vec{\nabla}\vec{\nabla}\mu:\vec{\nabla}\vec{V}+\vec{\nabla}\mu\cdot\vec{\nabla}\left(\vec{\nabla}\cdot\vec{V}\right)+\vec{\nabla}\vec{\nabla}\mu:\vec{\nabla}\vec{V}^{T}+\vec{\nabla}\mu\cdot\Delta\vec{V}.$$

If the fluid motion occurs in gravity force field, then there is potential $U=gz$, where $z$ is the vertical coordinate and the body force per unit mass is $\vec{f}=-\vec{\nabla}U$. In this case $\vec{\nabla}\cdot\rho\vec{f}=\vec{\nabla}\rho\cdot\vec{f}+\rho\vec{\nabla}\cdot\vec{f}=-\vec{\nabla}\rho\cdot\vec{\nabla}U-\rho\Delta U$.

Function $U$ is linear; therefore $\Delta U=0$ and

$$\vec{\nabla}\cdot\rho\vec{f}=-\vec{\nabla}\rho\cdot\vec{\nabla}U.$$

As a result of applied divergence operation to Navier–Stokes equation at $\mu\neq const$, we obtain scalar equation:

$$\rho\frac{d}{dt}\left(\vec{\nabla}\cdot\vec{V}\right)+\vec{\nabla}\rho\cdot\frac{d\vec{V}}{dt}+\rho\vec{\nabla}\vec{V}:\vec{\nabla}\vec{V}=-\Delta p'+\vec{\nabla}2\mu\cdot\left(\Delta V+\vec{\nabla}\left(\vec{\nabla}\cdot\vec{V}\right)\right)+$$
$$+2\mu\Delta\left(\vec{\nabla}\cdot\vec{V}\right)+\vec{\nabla}\vec{\nabla}:2\mu\underline{S}-\vec{\nabla}\rho\cdot\vec{\nabla}U. \quad (21)$$

In the case of incompressible fluid, we have

$$\rho\vec{\nabla}\vec{V}:\vec{\nabla}\vec{V}=-\Delta p+\vec{\nabla}\cdot 2\mu\Delta\vec{V}+\vec{\nabla}\vec{\nabla}:2\mu\underline{S}; \quad (22)$$

if also $\mu=const$, then

$$\rho\vec{\nabla}\vec{V}:\vec{\nabla}\vec{V}=-\Delta p. \quad (23)$$

Now we consider the general case of fluid motion, taking into account its compressibility.

The set of equations of motion of an incompressible fluid contains two—Navier–Stokes and continuity (one vector equation and one scalar equation) [2, 3]:

$$\begin{cases}\rho\dfrac{d\vec{V}}{dt}=-\vec{\nabla}p+\mu\Delta\vec{V}+\rho\vec{f}\\ \vec{\nabla}\cdot\vec{V}=0\end{cases}. \quad (24)$$

This set of two equations is closed: it contains two unknown quantities—velocity vector $\vec{V}$ and pressure per two equations. The set describes laminar flows; in turbulent flows it becomes unclosed because Reynolds stress tensor appeared.

In case of compressible flows at $\rho \neq const$, divergence of velocity is $\vec{\nabla} \cdot \vec{V} \neq 0$, and the Navier–Stokes equation (Eq. (19)) of a fluid motion at $\mu = const$ has the form:

$$\rho \frac{d\vec{V}}{dt} = -\vec{\nabla}p' + \mu\Delta\vec{V} + \mu\vec{\nabla}\left(\vec{\nabla} \cdot \vec{V}\right) + \rho\vec{f}, \tag{25}$$

where $p'$ is defined by Eq. (17). Continuity equation is written in the form of Eq. (1). If $\rho \neq const$, the set of equations (Eq. (25) and Eq. (1)) becomes unclosed, because density will also be unknown. To close the set of equation, energy equation is used, which contains one more unknown scalar quantity—temperature $T$. To determine temperature $T$, state equation is used; usually in fluid dynamics, it is the Mendeleev-Clapeyron equation. Energy equation could be written as the equation of specific internal energy transport:

$$\rho \frac{du}{dt} = -\vec{\nabla} \cdot \vec{q} + \underline{\sigma} : \vec{\nabla}\vec{V} + \rho q_s, \tag{26}$$

where $u$ is the specific internal energy (for ideal gas it could be expressed with the help of the isochore heat capacity, $du = c_v dT$); $\vec{q}$ is the heat flux vector (in laminar flow by Fourier's law, $\vec{q} = -\lambda\vec{\nabla}T$); $\lambda$ is the thermal conductivity of the material; and $q_s$ is the heat flux from internal or external sources.

Mendeleev-Clapeyron equation has the form

$$p = \rho RT, \tag{27}$$

where $R$ is the universal gas constant. In case of real gas or fluid, the state equation becomes more complicated.

For liquids it usually supposes $\rho = const$. This condition is applicable for gas motions and also in case the velocities of gas particles are less than 1/3 of sound velocity.

Eqs. (1), (25), (26), and (27) are valid for laminar regime of motion. In case of turbulent regime in these equations, correlations will appear, caused by velocity, density, and temperature pulsations. For closure of the set of equations of turbulent motion, additional relations are required.

## 4. Vorticity vector and its associated tensor

Vorticity $\vec{\omega}$ is a vector quantity, which characterizes velocity field:

$$\vec{\omega} = \frac{1}{2} rot\vec{V} = \frac{1}{2}\vec{\nabla} \times \vec{V}, \tag{28}$$

in component form

$$\vec{\omega} = \frac{1}{2}\frac{\partial V_j}{\partial x_i}\varepsilon_{kij}\vec{e}_k,$$

where $\varepsilon_{ijk}$ is the Levi-Civita tensor in component form. Components of $\vec{\omega}$ vector are

$$\left.\begin{aligned}\omega_1=\omega_x=\frac{1}{2}\left(\frac{\partial V_z}{\partial y}-\frac{\partial V_y}{\partial z}\right)=\frac{1}{2}\left(\frac{\partial V_3}{\partial x_2}-\frac{\partial V_2}{\partial x_3}\right)\\ \omega_2=\omega_y=\frac{1}{2}\left(\frac{\partial V_x}{\partial z}-\frac{\partial V_z}{\partial x}\right)=\frac{1}{2}\left(\frac{\partial V_1}{\partial x_3}-\frac{\partial V_3}{\partial x_1}\right)\\ \omega_3=\omega_z=\frac{1}{2}\left(\frac{\partial V_y}{\partial x}-\frac{\partial V_x}{\partial y}\right)=\frac{1}{2}\left(\frac{\partial V_2}{\partial x_1}-\frac{\partial V_1}{\partial x_2}\right)\end{aligned}\right\}. \tag{29}$$

Vorticity vector $\vec{\omega}$ and spin tensor $\underline{\Omega}$ are connected with each other through Levi-Civita tensor ${}^3\underline{\varepsilon}=\varepsilon_{ijk}\vec{e}_i\vec{e}_j\vec{e}_k$. These quantities are mutually associated. They say that tensor $\underline{\Omega}=\frac{1}{2}\left(\vec{\nabla}\vec{V}-\vec{\nabla}\vec{V}^T\right)$ is associated to vector $\vec{\omega}=\frac{1}{2}\vec{\nabla}\times\vec{V}$, because the following relations are satisfied:

$$\underline{\Omega}:\left({}^3\underline{\varepsilon}\right)=\left({}^3\underline{\varepsilon}\right):\underline{\Omega}=-2\vec{\omega}=-\vec{\nabla}\times\vec{V}. \tag{30}$$

And vice versa, vector $\vec{\omega}$ is associated with tensor $\underline{\Omega}$ since the following expression is valid:

$$\left({}^3\underline{\varepsilon}\right)\cdot\vec{\omega}=\vec{\omega}\cdot\left({}^3\underline{\varepsilon}\right)=\underline{\Omega}. \tag{31}$$

In Eq. (30) spin tensor $\underline{\Omega}$ is translated to the vector $\vec{\omega}$; in Eq. (31) vector $\vec{\omega}$ is associated with spin tensor.

Let us prove expression Eq. (30):

$$\begin{aligned}\underline{\Omega}:\left({}^3\underline{\varepsilon}\right)&=\frac{1}{2}\left(\vec{\nabla}\vec{V}-\vec{\nabla}\vec{V}^T\right):\left({}^3\underline{\varepsilon}\right)=\frac{1}{2}\left(\frac{\partial V_j}{\partial x_i}-\frac{\partial V_i}{\partial x_j}\right)\vec{e}_i\vec{e}_j:\varepsilon_{pqk}\vec{e}_p\vec{e}_q\vec{e}_k=\\ &=\frac{1}{2}\left(\frac{\partial V_j}{\partial x_i}-\frac{\partial V_i}{\partial x_j}\right)\varepsilon_{jik}\vec{e}_k=\frac{1}{2}\varepsilon_{ji1}\left(\frac{\partial V_j}{\partial x_i}-\frac{\partial V_i}{\partial x_j}\right)\vec{e}_1+\frac{1}{2}\varepsilon_{ji2}\left(\frac{\partial V_j}{\partial x_i}-\frac{\partial V_i}{\partial x_j}\right)\vec{e}_2+\\ &+\frac{1}{2}\varepsilon_{ji3}\left(\frac{\partial V_j}{\partial x_i}-\frac{\partial V_i}{\partial x_j}\right)\vec{e}_3=\frac{1}{2}\left(\frac{\partial V_2}{\partial x_3}-\frac{\partial V_3}{\partial x_2}-\frac{\partial V_3}{\partial x_2}+\frac{\partial V_2}{\partial x_3}\right)\vec{e}_1+\\ &+\frac{1}{2}\left(\frac{\partial V_3}{\partial x_1}-\frac{\partial V_1}{\partial x_3}-\frac{\partial V_1}{\partial x_3}+\frac{\partial V_3}{\partial x_1}\right)\vec{e}_2+\frac{1}{2}\left(\frac{\partial V_1}{\partial x_2}-\frac{\partial V_2}{\partial x_1}-\frac{\partial V_2}{\partial x_1}+\frac{\partial V_1}{\partial x_2}\right)\vec{e}_3=\\ &=-2\omega_1\vec{e}_1-2\omega_2\vec{e}_2-2\omega_3\vec{e}_3=-2\vec{\omega}=-\vec{\nabla}\times\vec{V}.\end{aligned}$$

The same for Eq. (31):

$$\left({}^3\underline{\varepsilon}\right)\cdot\vec{\omega}=\varepsilon_{ijk}\vec{e}_i\vec{e}_j\vec{e}_k\cdot\omega_s\vec{e}_s=\omega_k\varepsilon_{ijk}\vec{e}_i\vec{e}_j=\frac{1}{2}\frac{\partial V_s}{\partial x_t}\varepsilon_{kts}\varepsilon_{ijk}\vec{e}_i\vec{e}_j.$$

Let us descry components of this second-rank tensor: when $i=j$ they are equal to zero; when $i=1, j=2$ they are

$$\frac{1}{2}\frac{\partial V_s}{\partial x_t}\varepsilon_{kts}\varepsilon_{12k}=\frac{1}{2}\frac{\partial V_s}{\partial x_t}\varepsilon_{3ts}=\frac{1}{2}\left(\frac{\partial V_2}{\partial x_1}-\frac{\partial V_1}{\partial x_2}\right);$$

values for all $i,j=1,2,3$ could be obtained by the same way.

The matrix of components of this tensor is

$$\begin{pmatrix} 0 & \frac{1}{2}\left(\frac{\partial V_2}{\partial x_1}-\frac{\partial V_1}{\partial x_2}\right) & \frac{1}{2}\left(\frac{\partial V_3}{\partial x_1}-\frac{\partial V_1}{\partial x_3}\right) \\ \frac{1}{2}\left(\frac{\partial V_1}{\partial x_2}-\frac{\partial V_2}{\partial x_1}\right) & 0 & \frac{1}{2}\left(\frac{\partial V_2}{\partial x_1}-\frac{\partial V_1}{\partial x_2}\right) \\ \frac{1}{2}\left(\frac{\partial V_1}{\partial x_3}-\frac{\partial V_3}{\partial x_1}\right) & \frac{1}{2}\left(\frac{\partial V_2}{\partial x_3}-\frac{\partial V_3}{\partial x_2}\right) & 0 \end{pmatrix}.$$

It can be seen that it is a matrix of components of the antisymmetric tensor $\underline{\Omega}$, which means that relation (31) is valid.

It is easy to see also that

$$\vec{\omega}\cdot\underline{\Omega}=0, \vec{\omega}\times\underline{\Omega}=0, \vec{\nabla}\cdot\vec{\omega}=0. \tag{32}$$

## 5. Vorticity transport equation

Rotation of convective acceleration of a fluid particle could be written as

$$\vec{\nabla}\times\left[\left(\vec{V}\cdot\vec{\nabla}\right)\vec{V}\right]=2\left(\vec{\omega}\cdot\vec{\nabla}\right)\vec{V}-2\left(\vec{V}\cdot\vec{\nabla}\right)\vec{\omega}. \tag{33}$$

Evidence of this equation can be performed by writing convective acceleration according to the formula:

$$\left(\vec{V}\cdot\vec{\nabla}\right)\vec{V}=\vec{\nabla}\frac{\vec{V}^2}{2}-\vec{V}\times\left(\vec{\nabla}\times\vec{V}\right). \tag{34}$$

Gradient of vorticity is the pseudo tensor of rank 2:

$$\vec{\nabla}\vec{\omega}=\vec{\nabla}\left(\frac{1}{2}\vec{\nabla}\times\vec{V}\right)=\vec{e}_s\frac{\partial}{\partial x_s}\left(\frac{1}{2}\frac{\partial V_j}{\partial x_i}\varepsilon_{kij}\vec{e}_k\right)=\frac{1}{2}\frac{\partial^2 V_j}{\partial x_s\partial x_i}\varepsilon_{kij}\vec{e}_s\otimes\vec{e}_k. \tag{35}$$

Trace of this tensor is $tr\vec{\nabla}\vec{\omega}=0$. It is also possible to distinguish the symmetric and antisymmetric parts of this tensor:

$$\frac{1}{4}\left(\frac{\partial^2 V_j}{\partial x_s\partial x_i}\varepsilon_{kij}+\frac{\partial^2 V_j}{\partial x_k\partial x_i}\varepsilon_{sij}\right)\vec{e}_s\otimes\vec{e}_k, \frac{1}{4}\left(\frac{\partial^2 V_j}{\partial x_s\partial x_i}\varepsilon_{kij}-\frac{\partial^2 V_j}{\partial x_k\partial x_i}\varepsilon_{sij}\right)\vec{e}_s\otimes\vec{e}_k.$$

Let us assume that the fluid is incompressible, $\mu = const$, and its motion occurs in the field of potential mass forces. In this case with the help of Eq. (34), we can obtain the Navier–Stokes equation in the form:

$$\frac{\partial\vec{V}}{\partial t}+\vec{\nabla}\left(\frac{V^2}{2}+P+U\right)+\left(\vec{\nabla}\times\vec{V}\right)\times\vec{V}=\nu\Delta\vec{V};$$

now we apply curl operation ($rot=\vec{\nabla}\times$) to the left and right parts of this equation:

$$\frac{\partial}{\partial t}\left(\vec{\nabla}\times\vec{V}\right)+\vec{\nabla}\times\left[\left(\vec{\nabla}\times\vec{V}\right)\times\vec{V}\right]=\nu\Delta\left(\vec{\nabla}\times\vec{V}\right).$$

We could rewrite the second term of the left part in this form:

$$\left(\vec{V}\cdot\vec{\nabla}\right)\left(\vec{\nabla}\times\vec{V}\right)-\left[\left(\vec{\nabla}\times\vec{V}\right)\cdot\vec{\nabla}\right]\vec{V}+\left(\vec{\nabla}\times\vec{V}\right)\left(\vec{\nabla}\cdot\vec{V}\right)-\vec{V}\left(\vec{\nabla}\cdot 2\vec{\omega}\right),$$

but $\vec{\nabla}\cdot\vec{\omega}=0$, and fluid is incompressible ($\vec{\nabla}\cdot\vec{V}=0$); therefore, this term finally can be written as

$$2\left(\vec{V}\cdot\vec{\nabla}\right)\vec{\omega}-2\left(\vec{\omega}\cdot\vec{\nabla}\right)\vec{V}.$$

As a result, we can obtain transport equation of vortices in an incompressible viscous fluid, which is named as the generalized Helmholtz equation:

$$\frac{\partial\vec{\omega}}{\partial t}+\left(\vec{V}\cdot\vec{\nabla}\right)\vec{\omega}-\left(\vec{\omega}\cdot\vec{\nabla}\right)\vec{V}=\nu\Delta\vec{\omega}, \tag{36}$$

or in more compact form:

$$\frac{d\vec{\omega}}{dt}-\left(\vec{\omega}\cdot\vec{\nabla}\right)\vec{V}=\nu\Delta\vec{\omega}. \tag{37}$$

It is necessary to note that in the case of compressible fluid and at $\mu\neq const$, this equation becomes much more complicated.

This transport equation can be written in another form, considering the equality $\left(\vec{V}\cdot\vec{\nabla}\right)\vec{\omega}-\left(\vec{\omega}\cdot\vec{\nabla}\right)\vec{V}=\vec{\nabla}\times\left(\vec{\omega}\times\vec{V}\right)$. Then

$$\frac{\partial\vec{\omega}}{\partial t}+\vec{\nabla}\times\left(\vec{\omega}\times\vec{V}\right)=\nu\Delta\vec{\omega}. \tag{38}$$

If we apply divergence operation to Eq. (36), then for incompressible fluid we obtain

$$\vec{\nabla}\cdot\left[\frac{\partial\vec{\omega}}{\partial t}+\left(\vec{V}\cdot\vec{\nabla}\right)\vec{\omega}-\left(\vec{\omega}\cdot\vec{\nabla}\right)\vec{V}\right]=\vec{\nabla}\cdot\nu\Delta\vec{\omega};$$

because $\vec{\nabla}\cdot\vec{\omega}=0$, we have

$$\vec{\nabla}\cdot\left[\left(\vec{V}\cdot\vec{\nabla}\right)\vec{\omega}\right]-\vec{\nabla}\left[\left(\vec{\omega}\cdot\vec{\nabla}\right)\vec{V}\right]=0.$$

On the other hand,

$$\begin{aligned}\vec{\nabla}\cdot\left[\left(\vec{V}\cdot\vec{\nabla}\right)\vec{\omega}\right]&=\vec{\nabla}\vec{V}:\vec{\nabla}\vec{\omega}+\left(\vec{V}\cdot\vec{\nabla}\right)\left(\vec{\nabla}\cdot\vec{\omega}\right);\ \vec{\nabla}\cdot\left[\left(\vec{\omega}\cdot\vec{\nabla}\right)\vec{V}\right]\\&=\vec{\nabla}\vec{V}:\vec{\nabla}\vec{\omega}+\left(\vec{\omega}\cdot\vec{\nabla}\right)\left(\vec{\nabla}\cdot\vec{V}\right),\end{aligned}$$

and, finally, we have $0=0$, i.e., we shall not obtain a new expression.

For the second power of vorticity, we can write

$$\omega^2=\vec{\nabla}\vec{V}^T:\vec{\nabla}\vec{V}-\vec{\nabla}\vec{V}:\vec{\nabla}\vec{V}, \tag{39}$$

and now, if we scalar multiply transport equation of vortices by $\vec{\omega}$:

$$\vec{\omega}\cdot\left[\frac{\partial\vec{\omega}}{\partial t}+\left(\vec{V}\cdot\vec{\nabla}\right)\vec{\omega}-\left(\vec{\omega}\cdot\vec{\nabla}\right)\vec{V}\right]=\vec{\omega}\cdot\nu\Delta\vec{\omega};$$

then we obtain scalar transport equation of $\omega^2$:

$$\frac{d\omega^2}{dt}-\left(\vec{\omega}\otimes\vec{\omega}\right):\vec{\nabla}\vec{V}=\nu\Delta\omega^2-\nu\vec{\nabla}\vec{\omega}^T:\vec{\nabla}\vec{\omega}. \tag{40}$$

For incompressible fluid $\vec{\nabla}\cdot\underline{S}=\vec{\nabla}\cdot\underline{\Omega}$ because $\vec{\nabla}\cdot\underline{S}=\frac{1}{2}\Delta\vec{V}+\frac{1}{2}\vec{\nabla}\left(\vec{\nabla}\cdot\vec{V}\right)$ and $\vec{\nabla}\cdot\underline{\Omega}=\frac{1}{2}\Delta\vec{V}-\frac{1}{2}\vec{\nabla}\left(\vec{\nabla}\cdot\vec{V}\right)$.

As we already know $\vec{V}\times\left(\vec{\nabla}\times\vec{V}\right)=-2\vec{V}\cdot\underline{\Omega}$; therefore, Eq. (34) can be written as

$$\left(\vec{V}\cdot\vec{\nabla}\right)\vec{V}=\vec{\nabla}\frac{V^2}{2}+2\vec{V}\cdot\underline{\Omega}. \tag{41}$$

Let us write one more equation:

$$\vec{\nabla}\cdot\left[\vec{V}\times\left(\vec{\nabla}\times\vec{V}\right)\right]=-\vec{\nabla}\cdot\left[\vec{V}\cdot2\underline{\Omega}\right]=\vec{\nabla}\cdot\left[\vec{V}\cdot\vec{\nabla}\vec{V}^T-\vec{V}\cdot\vec{\nabla}\vec{V}\right]=$$

$$=\vec{\nabla}\cdot\left[V_i\frac{\partial V_i}{\partial x_j}-V_i\frac{\partial V_j}{\partial x_i}\right]\vec{e_j}=\frac{\partial}{\partial x_j}\left[V_i\frac{\partial V_i}{\partial x_j}-V_i\frac{\partial V_j}{\partial x_i}\right]=\frac{\partial V_i}{\partial x_j}\frac{\partial V_i}{\partial x_j}+V_i\frac{\partial^2 V_i}{\partial x_j\partial x_j}-$$

$$-\frac{\partial V_i}{\partial x_j}\frac{\partial V_j}{\partial x_i}-V_i\frac{\partial^2 V_j}{\partial x_i\partial x_j}=\vec{\nabla}\vec{V}^T:\vec{\nabla}\vec{V}-\vec{\nabla}\vec{V}:\vec{\nabla}\vec{V}+\vec{V}\cdot\Delta\vec{V}-\vec{V}\cdot\vec{\nabla}\left(\vec{\nabla}\cdot\vec{V}\right).$$

Therefore,

$$\vec{\nabla}\cdot\left[\vec{V}\times\left(\vec{\nabla}\times\vec{V}\right)\right]=\vec{\nabla}\vec{V}^T:\vec{\nabla}\vec{V}-\vec{\nabla}\vec{V}:\vec{\nabla}\vec{V}+\vec{V}\cdot\left[\Delta\vec{V}-\vec{\nabla}\left(\vec{\nabla}\cdot\vec{V}\right)\right] \tag{42}$$

This equation also can be written in the next form:

$$\vec{\nabla}\cdot\left[\vec{V}\times\left(\vec{\nabla}\times\vec{V}\right)\right]=2\underline{\Omega}:\vec{\nabla}\vec{V}+2\vec{V}\cdot\left(\vec{\nabla}\cdot\underline{\Omega}\right)$$

or

$$\vec{\nabla}\cdot\left[\vec{V}\times\left(\vec{\nabla}\times\vec{V}\right)\right]=2\underline{\Omega}:\underline{\Omega}+2\underline{\Omega}:\underline{S}+2\vec{V}\cdot\left(\vec{\nabla}\cdot\underline{\Omega}\right).$$

One more interesting relation is

$$\vec{V}\cdot\vec{\nabla}\vec{V}=\vec{V}\cdot\underline{\Omega}+\vec{V}\cdot\underline{S},$$

but as we already mentioned, $\vec{V}\times\left(\vec{\nabla}\times\vec{V}\right)=2\vec{V}\cdot\underline{\Omega}$; therefore,

$$\vec{V}\cdot\underline{S}=\vec{\nabla}\frac{V^2}{2}-\frac{1}{2}\vec{V}\times\left(\vec{\nabla}\times\vec{V}\right). \tag{43}$$

The vector product of gradients of scalar functions gives us a vector; in terms of rotation of a vector function, we can write

$$\vec{\nabla}f \times \vec{\nabla}\varphi = \vec{\nabla} \times \left(f\vec{\nabla}\varphi\right). \tag{44}$$

Really, the left part of this equation is a vector:

$$\vec{\nabla}f \times \vec{\nabla}\varphi = \left(\frac{\partial f}{\partial x_2}\frac{\partial \varphi}{\partial x_3} - \frac{\partial f}{\partial x_3}\frac{\partial \varphi}{\partial x_2}\right)\vec{e}_1 + \left(\frac{\partial f}{\partial x_3}\frac{\partial \varphi}{\partial x_1} - \frac{\partial f}{\partial x_1}\frac{\partial \varphi}{\partial x_3}\right)\vec{e}_2 + \left(\frac{\partial f}{\partial x_1}\frac{\partial \varphi}{\partial x_2} - \frac{\partial f}{\partial x_2}\frac{\partial \varphi}{\partial x_1}\right)\vec{e}_3,$$

and the right part is.

$$\begin{aligned}\vec{\nabla} \times \left(f\vec{\nabla}\varphi\right) &= \vec{e}_i\frac{\partial}{\partial x_i} \times f\frac{\partial \varphi}{\partial x_j}\vec{e}_j = \left(\vec{e}_i \times \vec{e}_j\right)\frac{\partial}{\partial x_i}f\frac{\partial \varphi}{\partial x_j} = \varepsilon_{kij}\frac{\partial}{\partial x_i}\left(f\frac{\partial \varphi}{\partial x_j}\right)\vec{e}_k = \\ &= \varepsilon_{1ij}\frac{\partial}{\partial x_i}\left(f\frac{\partial \varphi}{\partial x_j}\right)\vec{e}_1 + \varepsilon_{2ij}\frac{\partial}{\partial x_i}\left(f\frac{\partial \varphi}{\partial x_j}\right)\vec{e}_2 + \varepsilon_{3ij}\frac{\partial}{\partial x_i}\left(f\frac{\partial \varphi}{\partial x_j}\right)\vec{e}_3 = \\ &= \left(\frac{\partial}{\partial x_2}f\frac{\partial \varphi}{\partial x_3} - \frac{\partial}{\partial x_3}f\frac{\partial \varphi}{\partial x_2}\right)\vec{e}_1 + \left(\frac{\partial}{\partial x_3}f\frac{\partial \varphi}{\partial x_1} - \frac{\partial}{\partial x_1}f\frac{\partial \varphi}{\partial x_3}\right)\vec{e}_2 \\ &\quad + \left(\frac{\partial}{\partial x_1}f\frac{\partial \varphi}{\partial x_2} - \frac{\partial}{\partial x_2}f\frac{\partial \varphi}{\partial x_1}\right)\vec{e}_3 = \\ &= \left(\frac{\partial f}{\partial x_2}\frac{\partial \varphi}{\partial x_3} - \frac{\partial f}{\partial x_3}\frac{\partial \varphi}{\partial x_2}\right)\vec{e}_1 + \left(\frac{\partial f}{\partial x_3}\frac{\partial \varphi}{\partial x_1} - \frac{\partial f}{\partial x_1}\frac{\partial \varphi}{\partial x_3}\right)\vec{e}_2 + \left(\frac{\partial f}{\partial x_1}\frac{\partial \varphi}{\partial x_2} - \frac{\partial f}{\partial x_2}\frac{\partial \varphi}{\partial x_1}\right)\vec{e}_3.\end{aligned}$$

Therefore, Eq. (44) is valid.

The vector product of gradients of scalar functions also can be written as

$$\vec{\nabla}f \times \vec{\nabla}\varphi = \frac{1}{2}\left(\vec{\nabla}\varphi \otimes \vec{\nabla}f - \vec{\nabla}f \otimes \vec{\nabla}\varphi\right) : \left({}^3\underline{\varepsilon}\right) = -\vec{\nabla} \cdot \left[\left(f\vec{\nabla}\varphi\right) \cdot \left({}^3\underline{\varepsilon}\right)\right]. \tag{45}$$

Here we have in component form:

$$\begin{aligned}\vec{\nabla} \cdot \left[\left(f\vec{\nabla}\varphi\right) \cdot {}^3\underline{\varepsilon}\right] &= \vec{\nabla} \cdot \left[f\frac{\partial \varphi}{\partial x_s}\vec{e}_s \cdot \varepsilon_{kij}\vec{e}_k\vec{e}_i\vec{e}_j\right] = \vec{\nabla} \cdot \left[f\frac{\partial \varphi}{\partial x_k}\varepsilon_{kij}\vec{e}_i\vec{e}_j\right] = \\ &= \vec{e}_t\frac{\partial}{\partial x_t} \cdot f\frac{\partial \varphi}{\partial x_k}\varepsilon_{kij}\vec{e}_i\vec{e}_j = \frac{\partial}{\partial x_i}\left(f\frac{\partial \varphi}{\partial x_k}\right)\varepsilon_{kij}\vec{e}_j = \\ &= \left(\frac{\partial}{\partial x_i}f\frac{\partial \varphi}{\partial x_k}\right)\varepsilon_{ki1}\vec{e}_1 + \left(\frac{\partial}{\partial x_i}f\frac{\partial \varphi}{\partial x_k}\right)\varepsilon_{ki2}\vec{e}_2 + \left(\frac{\partial}{\partial x_i}f\frac{\partial \varphi}{\partial x_k}\right)\varepsilon_{ki3}\vec{e}_3 = \\ &= \left(\frac{\partial f}{\partial x_3} \cdot \frac{\partial \varphi}{\partial x_2} - \frac{\partial f}{\partial x_2} \cdot \frac{\partial \varphi}{\partial x_3}\right)\vec{e}_1 + \left(\frac{\partial f}{\partial x_1} \cdot \frac{\partial \varphi}{\partial x_3} - \frac{\partial f}{\partial x_3} \cdot \frac{\partial \varphi}{\partial x_1}\right)\vec{e}_2 \\ &\quad + \left(\frac{\partial f}{\partial x_2} \cdot \frac{\partial \varphi}{\partial x_1} - \frac{\partial f}{\partial x_1} \cdot \frac{\partial \varphi}{\partial x_2}\right)\vec{e}_3.\end{aligned}$$

It is easy to see that this is equal to expression for $\vec{\nabla}f \times \vec{\nabla}\varphi$ with the minus sign.

## 6. Mechanical energy equation

Mechanical energy balance equation can be obtained as a scalar product of each member of Eq. (12) on velocity vector $\vec{V}$:

$$\vec{V} \cdot \rho \left[ \frac{\partial \vec{V}}{\partial t} + \left( \vec{V} \cdot \vec{\nabla} \right) \vec{V} \right] = \vec{V} \cdot \left( \vec{\nabla} \cdot \underline{\sigma} \right) + \vec{V} \cdot \rho \vec{f}. \tag{46}$$

Transformations of the left part lead us to the following results:

$$\vec{V} \cdot \frac{\partial \vec{V}}{\partial t} = V_k \vec{e}_k \cdot \frac{\partial}{\partial t} V_i \vec{e}_i = V_k \frac{\partial V_i}{\partial t} \left( \vec{e}_k \cdot \vec{e}_i \right) = V_k \frac{\partial V_i}{\partial t} \delta_{ki} = V_k \frac{\partial V_k}{\partial t} = \frac{1}{2} \frac{\partial}{\partial t} V_k V_k = \frac{1}{2} \frac{\partial V^2}{\partial t};$$

$$V \cdot \left[ \left( \vec{V} \cdot \vec{\nabla} \right) \vec{V} \right] = V_k \vec{e}_k \cdot V_j \frac{\partial}{\partial x_j} V_i \vec{e}_i = V_k V_j \frac{\partial V_i}{\partial x_j} \left( \vec{e}_k \cdot \vec{e}_i \right) =$$

$$= V_k V_j \frac{\partial V_i}{\partial x_j} \delta_{ki} = V_k V_j \frac{\partial V_k}{\partial x_j} = V_j \frac{\partial}{\partial x_j} \frac{1}{2} V_k V_k = \left( \vec{V} \cdot \vec{\nabla} \right) \frac{V^2}{2}.$$

It is easy to see that in sum the left part is the material derivative of kinetic energy of a fluid particle—quantity $\frac{d}{dt} \frac{V^2}{2}$. Then Eq. (46) takes the form:

$$\rho \frac{d}{dt} \left[ \frac{V^2}{2} \right] = \vec{V} \cdot \left( \vec{\nabla} \cdot \underline{\sigma} \right) + \vec{V} \cdot \rho \vec{f}. \tag{47}$$

Usually stress tensor is defined as the sum $\underline{\sigma} = -p\underline{E} + \underline{\tau}$, where $\underline{\tau}$ is the shear stress tensor. Then the first term of the right part of Eq. (46) takes the form:

$$\vec{V} \cdot \left( \vec{\nabla} \cdot \underline{\sigma} \right) = -\vec{V} \cdot \left( \vec{\nabla} \cdot p\underline{E} \right) + \vec{V} \cdot \left( \vec{\nabla} \cdot \underline{\tau} \right).$$

Now we can represent equation of mechanical energy balance (Eq. (46)) considering $\vec{\nabla} \cdot p\underline{E} = \vec{\nabla} p$ as follows:

$$\rho \frac{d}{dt} \left[ \frac{V^2}{2} \right] = -\vec{V} \cdot \vec{\nabla} p + \vec{V} \cdot \left( \vec{\nabla} \cdot \underline{\tau} \right) + \rho \vec{V} \cdot \vec{f}. \tag{48}$$

The first member of the right part of Eq. (46) is power of stresses $\vec{V} \cdot \left( \vec{\nabla} \cdot \underline{\sigma} \right)$, which can be written in the form:

$$\vec{V} \cdot \left( \vec{\nabla} \cdot \underline{\sigma} \right) = \vec{\nabla} \cdot \left( \vec{V} \cdot \underline{\sigma} \right) - \underline{\sigma} : \vec{\nabla} \vec{V}. \tag{49}$$

It is easy to be proven if we rewrite this expression in component form in Cartesian coordinates. In this case the left part of Eq. (49) is

$$\vec{V} \cdot \left( \vec{\nabla} \cdot \underline{\sigma} \right) = V_k \vec{e}_k \cdot \left( \vec{e}_i \frac{\partial}{\partial x_i} \cdot \sigma_{sj} \vec{e}_s \vec{e}_j \right) = V_k \vec{e}_k \cdot \left[ \frac{\partial \sigma_{sj}}{\partial x_i} \left( \vec{e}_i \cdot \vec{e}_s \right) \vec{e}_j \right] =$$

$$= V_k \vec{e}_k \cdot \frac{\partial \sigma_{ij}}{\partial x_i} \vec{e}_j = V_k \frac{\partial \sigma_{ij}}{\partial x_i} \left( \vec{e}_k \cdot \vec{e}_j \right) = V_k \frac{\partial \sigma_{ik}}{\partial x_i}.$$

The first term of the right part of Eq. (49) in component form is

$$\vec{\nabla} \cdot \left( \vec{V} \cdot \underline{\sigma} \right) = \vec{\nabla} \cdot \left( V_s \vec{e}_s \cdot \sigma_{ij} \vec{e}_i \vec{e}_j \right) = \vec{\nabla} \cdot \left[ V_s \sigma_{ij} \left( \vec{e}_s \cdot \vec{e}_i \right) \vec{e}_j \right] = \vec{\nabla} \cdot \left( V_i \sigma_{ij} \vec{e}_j \right) =$$

$$= \vec{e}_k \frac{\partial}{\partial x_k} \cdot V_i \sigma_{ij} \vec{e}_j = \delta_{kj} \frac{\partial}{\partial x_k} V_i \sigma_{ij} = \frac{\partial}{\partial x_k} V_i \sigma_{ik} = \frac{\partial V_i}{\partial x_k} \sigma_{ik} + V_i \frac{\partial \sigma_{ik}}{\partial x_k}.$$

The second term of the right part of Eq. (49) in component form is

$$\sigma : \vec{\nabla}\vec{V} = \sigma_{ij}\vec{e}_i\vec{e}_j : \frac{\partial V_s}{\partial x_k}\vec{e}_k\vec{e}_s = \sigma_{ij}\frac{\partial V_s}{\partial x_k}\left(\vec{e}_j \cdot \vec{e}_k\right)\left(\vec{e}_i \cdot \vec{e}_s\right) =$$

$$= \sigma_{ij}\frac{\partial V_s}{\partial x_k}\delta_{jk}\delta_{is} = \sigma_{ij}\frac{\partial V_i}{\partial x_j} = \frac{\partial V_i}{\partial x_k}\sigma_{ik}.$$

Finally Eq. (49) in component form is

$$V_k\frac{\partial \sigma_{ik}}{\partial x_i} = \frac{\partial V_i}{\partial x_k}\sigma_{ik} + V_i\frac{\partial \sigma_{ik}}{\partial x_k} - \frac{\partial V_i}{\partial x_k}\sigma_{ik}.$$

Due to the symmetry of the stress tensor $\underline{\sigma} = \underline{\sigma}^T$, this expression is an identity (i.e., the left side is equal to the right one). Indeed, the second term of the right-hand side of this relation, after re-designating the index $i$ by $k$ and vice versa, takes the form $V_k\frac{\partial \sigma_{ki}}{\partial x_i}$, but $\sigma_{ki} = \sigma_{ik}$; therefore both parts of the expression are equal to each other. Thus, equality (49) is valid.

We could simplify the last term of the right part of Eq. (46) if we introduce potential $U$ of mass forces in field of gravity ($z$ axis is positive upwards) as earlier $\vec{f} = -\vec{\nabla}U$, $U = gz = \vec{r} \cdot \vec{g}$, where $\vec{g}$ is the gravity acceleration vector in the field of gravity. Then

$$\vec{V} \cdot \rho\vec{f} = -\rho\vec{V} \cdot \vec{\nabla}U = -\rho\frac{d\vec{r}}{dt} \cdot \left(\vec{\nabla}U\right) = -\rho\frac{1}{dt}\left(\vec{\nabla}U \cdot d\vec{r}\right) = -\rho\frac{dU}{dt}. \tag{50}$$

After substituting the above expressions into Eq. (46), we obtain the equation for mechanical energy of a fluid flow:

$$\rho\frac{d}{dt}\left(\frac{V^2}{2} + U\right) = \vec{\nabla} \cdot \left(\vec{V} \cdot \underline{\sigma}\right) - \underline{\sigma} : \vec{\nabla}\vec{V}. \tag{51}$$

In the left part of this equation, we observe the total mechanical energy of a fluid flow as the sum of kinetic and potential energy of the flow [6]. Often the right part of Eq. (51) is written in another form, where stress tensor is written as the sum $\sigma = -p\underline{E} + \underline{\tau}$. Then the right part of Eq. (51) will have the form:

$$-\vec{\nabla} \cdot \left(\vec{V} \cdot p\underline{E}\right) + \vec{\nabla} \cdot \left(\vec{V} \cdot \underline{\tau}\right) + p\underline{E} : \vec{\nabla}\vec{V} - \underline{\tau} : \vec{\nabla}\vec{V}. \tag{52}$$

The first member is

$$-\vec{\nabla} \cdot \left(\vec{V} \cdot p\underline{E}\right) = -\vec{\nabla} \cdot \left(V_i\vec{e}_i \cdot p\delta_{jk}\vec{e}_j\vec{e}_k\right) = -\vec{\nabla} \cdot \left(\vec{V}_i p\delta_{jk}\delta_{ij}\vec{e}_k\right) =$$

$$= -\vec{\nabla} \cdot \left(pV_k\vec{e}_k\right) = -\vec{e}_s\frac{\partial}{\partial x_s} \cdot pV_k\vec{e}_k = -\frac{\partial pV_k}{\partial x_s}\delta_{sk} = -\frac{\partial}{\partial x_k}pV_k =$$

$$= -V_k\frac{\partial p}{\partial x_k} - p\frac{\partial V_k}{\partial x_k} = -\left(\vec{V} \cdot \vec{\nabla}\right)p - p\vec{\nabla} \cdot \vec{V}.$$

The third member is

$$p\underline{E} : \vec{\nabla}\vec{V} = p\delta_{ij}\vec{e}_i\vec{e}_j : \frac{\partial V_s}{\partial x_k}\vec{e}_k\vec{e}_s = p\frac{\partial V_s}{\partial x_k}\delta_{ij}\delta_{jk}\delta_{is} = p\frac{\partial V_k}{\partial x_k} = p\vec{\nabla} \cdot \vec{V}.$$

When we substitute these terms in Eq. (52) and then in Eq. (51), the equation for mechanical energy of a fluid flow will take the form:

$$\rho \frac{d}{dt}\left(\frac{V^2}{2}+U\right) = -\left(\vec{V}\cdot\vec{\nabla}\right)p + \vec{\nabla}\left(\vec{V}\cdot\underline{\tau}\right) - \tau : \vec{\nabla}\vec{V}. \tag{53}$$

As a result, we could conclude that the rate of change of total mechanical energy of a flow is equal to the sum of the powers of the pressure forces and viscous friction.

Navier–Stokes equation for a steady flow of viscous incompressible fluid is

$$\rho\left(\vec{V}\cdot\vec{\nabla}\right)\vec{V} = -\vec{\nabla}p + \mu\Delta\vec{V} + \rho\vec{f}. \tag{54}$$

The Laplacian of velocity in the right part can be written in the form:

$$\Delta\vec{V} = \vec{\nabla}\left(\vec{\nabla}\cdot\vec{V}\right) - \vec{\nabla}\times\left(\vec{\nabla}\times\vec{V}\right). \tag{55}$$

It can be obtained by consideration of operation $\vec{\nabla}\times\left(\vec{\nabla}\times\vec{V}\right)$:

$$\vec{\nabla}\times\left(\vec{\nabla}\times\vec{V}\right) = \vec{e}_i\frac{\partial}{\partial x_i}\times\left(\frac{\partial V_j}{\partial x_k}\varepsilon_{skj}\vec{e}_s\right) =$$

$$= \frac{\partial^2 V_j}{\partial x_i \partial x_k}\varepsilon_{skj}\varepsilon_{lis}\vec{e}_l = \frac{\partial^2 V_j}{\partial x_i \partial x_k}\varepsilon_{skj}\varepsilon_{1is}\vec{e}_1 + \frac{\partial^2 V_j}{\partial x_i \partial x_k}\varepsilon_{skj}\varepsilon_{2is}\vec{e}_2 + \frac{\partial^2 V_j}{\partial x_i \partial x_k}\varepsilon_{skj}\varepsilon_{3is}\vec{e}_3.$$

The member with the basis vector $\vec{e}_1$ is determined as

$$\frac{\partial^2 V_j}{\partial x_i \partial x_k}\varepsilon_{skj}\varepsilon_{1is} = \frac{\partial^2 V_j}{\partial x_2 \partial x_k}\varepsilon_{3kj} - \frac{\partial^2 V_j}{\partial x_3}\varepsilon_{2kj} = \frac{\partial^2 V_2}{\partial x_2 \partial x_1} - \frac{\partial^2 V_1}{\partial x_3 \partial x_3} + \frac{\partial^2 V_3}{\partial x_3 \partial x_1} =$$

$$= \frac{\partial}{\partial x_1}\left(\frac{\partial^2 V_1}{\partial x_1} + \frac{\partial^2 V_2}{\partial x_2} + \frac{\partial^2 V_3}{\partial x_3}\right) - \frac{\partial^2 V_1}{\partial x_1^2} - \frac{\partial^2 V_1}{\partial x_2^2} - \frac{\partial^2 V_1}{\partial x_3^2} = \frac{\partial}{\partial x_1}\left(\vec{\nabla}\cdot\vec{V}\right) - \Delta V_1.$$

The same can be written for the members with basis vectors $\vec{e}_2$ and $\vec{e}_3$. As a result, we obtain

$$\vec{\nabla}\times\left(\vec{\nabla}\times\vec{V}\right) = \left[\frac{\partial}{\partial x_1}\left(\vec{\nabla}\cdot\vec{V}\right) - \Delta V_1\right]\vec{e}_1 + \left[\frac{\partial}{\partial x_2}\left(\vec{\nabla}\cdot\vec{V}\right) - \Delta V_2\right]\vec{e}_2 +$$

$$+\left[\frac{\partial}{\partial x_3}\left(\vec{\nabla}\cdot\vec{V}\right) - \Delta V_3\right]\vec{e}_3 = \vec{\nabla}\left(\vec{\nabla}\cdot\vec{V}\right) - \Delta\vec{V}.$$

And therefore formula Eq. (55) is valid.

One more useful expression based on Eq. (55) is

$$\vec{\nabla}\cdot\underline{S} = \Delta\vec{V} + \vec{\nabla}\left(\vec{\nabla}\cdot\vec{V}\right) = 2\vec{\nabla}\left(\vec{\nabla}\cdot\vec{V}\right) - \vec{\nabla}\times\vec{\nabla}\times\vec{V}.$$

Using Eqs. (34) and (55) and considering the mass force in field of gravity with the help of potential $U$, $U = -gz$ ($z$ axis is positive upwards), we obtain instead Eq. (54):

$$\vec{\nabla}\frac{V^2}{2} - \vec{V} \times \left(\vec{\nabla} \times \vec{V}\right) = -\frac{1}{\rho}\vec{\nabla}p + \nu\vec{\nabla}\left(\vec{\nabla} \cdot \vec{V}\right) - \nu\vec{\nabla} \times \left(\vec{\nabla} \times \vec{V}\right) + \vec{\nabla}U,$$

where $\nu = \frac{\mu}{\rho}$ is the kinematic (momentum) viscosity of fluid.

In incompressible fluid $\rho = const$, $\vec{\nabla} \cdot \vec{V} = 0$ and then we have

$$\vec{\nabla}\left(\frac{V^2}{2} + \frac{p}{\rho} + gz\right) = \underbrace{\vec{V} \times \left(\vec{\nabla} \times \vec{V}\right) - \nu\vec{\nabla} \times \vec{\nabla} \times \vec{V}}_{\left(\vec{V} - \nu\vec{\nabla}\right) \times \vec{\nabla} \times \vec{V}}. \tag{56}$$

The gradient of total mechanical energy of a fluid particle $E = \frac{V^2}{2} + \frac{p}{\rho} + gz$ depends on vortex structure of the flow. When $\nabla \times \vec{V} = 0$ the right part of Eq. (56) is zero, and then $E = const$ in the whole area of the flow.

It is possible to obtain the divergence form of Eq. (54) for incompressible fluid considering $\vec{\nabla} \cdot \vec{V} = 0$ and using Eq. (55) and the relation:

$$\vec{\nabla} \cdot \left(\vec{V} \otimes \vec{V}\right) = \vec{V}\left(\vec{\nabla} \cdot \vec{V}\right) + \left(\vec{V} \cdot \vec{\nabla}\right)\vec{V}. \tag{57}$$

Then Eq. (54) will take the form:

$$\vec{\nabla} \cdot \left(\rho\vec{V} \otimes \vec{V}\right) = -\vec{\nabla}p + \vec{\nabla}\rho U - \mu\vec{\nabla} \times \vec{\nabla} \times \vec{V}. \tag{58}$$

Using concepts of identity tensor $\underline{I}$ and Levi-Civita tensor ${}^3\underline{\varepsilon} = \varepsilon_{ijk}\vec{e}_i\vec{e}_j\vec{e}_k$, we can write members of the right part in the divergence form, and as a result, the whole Navier–Stokes equation for steady flow of incompressible fluid can be written as

$$\vec{\nabla} \cdot \left[\rho\vec{V} \otimes \vec{V} + p\underline{I} - \rho U\underline{I} + \rho\nu\left({}^3\varepsilon\right) \cdot \left(\vec{\nabla} \times \vec{V}\right)\right] = 0. \tag{59}$$

The last term of Eq. (59) can be considered in a more simple form due to relation for incompressible fluid $\Delta\vec{V} = \vec{\nabla} \cdot \vec{\nabla}\vec{V} = \vec{\nabla} \cdot \underline{S}$. Finally, the divergence form of the Navier–Stokes equation for steady flow of an incompressible fluid is

$$\vec{\nabla} \cdot \left[\rho\left(\vec{V} \otimes \vec{V}\right) + (p + \rho gz)\underline{I} - \mu\underline{S}\right] = 0. \tag{60}$$

## 7. Energy equation for moving fluid

The first law of thermodynamics connects internal energy, heat, and work. In the case of moving fluid, it can be written as follows:

$$\rho\frac{du}{dt} = -\vec{\nabla} \cdot \vec{q} + \underline{\sigma} : \vec{\nabla}\vec{V} + q_v, \tag{61}$$

where $t$ is the time; $u$ is the specific internal energy; $\vec{q}$ is the heat flux density vector due to thermal conductivity; $\underline{\sigma}$ is the stress tensor; and $q_v$ is the value of heat entering into the particle volume from action of external or internal sources per unit time. In this expression, the colon denotes double scalar product of tensors; in this case these are the stress tensor and velocity gradient tensor.

The physical meaning of this equation is that the rate of change of internal energy per unit volume is equal to rate of energy supply due to heat conduction, due to dissipation of mechanical energy of the flow, and due to heat from external or internal sources. Since stress tensor $\underline{\sigma}$ can be written as $\underline{\sigma} = -p\underline{E} + \underline{\tau}$, where $\underline{\tau}$ is the shear stress tensor, taking into account material derivative definition, we can rewrite Eq. (61) in this form:

$$\rho\left[\frac{\partial u}{\partial t} + \left(\vec{V}\cdot\vec{\nabla}\right)u\right] = -\vec{\nabla}\cdot\vec{q} - p\vec{\nabla}\cdot\vec{V} + \underline{\tau} : \vec{\nabla}\vec{V} + q_v. \tag{62}$$

This is the energy equation in terms of transfer of specific internal energy $u$. Vector $\vec{q}$ in the energy equation is determined by Fourier's law:

$$\vec{q} = -\lambda\vec{\nabla}T, \tag{63}$$

where $T$ is the temperature and $\lambda$ is the coefficient of thermal conductivity.

Fourier's law of thermal conductivity can also be written in terms of enthalpy, which for an ideal gas is related to temperature by the formula $h = c_p T$, where $c_p$ is the isobar heat capacity. Then considering $\lambda = \rho c_p a$, where a is the thermal diffusivity, heat flux density vector can be written in the form:

$$\vec{q} = \lambda\vec{\nabla}T = \rho c_p a\vec{\nabla}T = \rho\frac{\nu}{a/\nu}\vec{\nabla}c_p T = \rho\frac{\nu}{\mathrm{Pr}}\vec{\nabla}h = \frac{\mu}{\mathrm{Pr}}\vec{\nabla}h,$$

where $\mathrm{Pr} = a/\nu$ is the Prandtl number.

In Cartesian coordinates Eq. (62) can be written as follows:

$$\rho\left[\frac{\partial u}{\partial t} + V_j\frac{\partial u}{\partial x_j}\right] = \frac{\partial}{\partial x_j}\lambda\frac{\partial T}{\partial x_j} - p\frac{\partial V_j}{\partial x_j} + \tau_{ij}\frac{\partial V_i}{\partial x_j} + q_v. \tag{64}$$

The terms $p\vec{\nabla}\cdot\vec{V}$ and $\underline{\tau} : \vec{\nabla}\vec{V}$ show us in a moving fluid heating or cooling can occur. The term $p\vec{\nabla}\cdot\vec{V}$ may cause significant change of temperature, when gas expands (compresses) rapidly. The term $\underline{\tau} : \vec{\nabla}\vec{V}$ is always positive; it characterizes dissipation of mechanical energy and its transformation to heat energy. This scalar quantity is usually named as Rayleigh dissipation function [6] and denoted as $\underline{\tau} : \vec{\nabla}\vec{V} = \Phi$. Let us write this function in Cartesian coordinates for Newtonian viscous fluid, when rheological relation has the form:

$$\underline{\tau} = -\frac{2}{3}\mu\left(\vec{\nabla}\cdot\vec{V}\right)\underline{E} + 2\mu\underline{S}, \tag{65}$$

where $\mu$ is the fluid shear viscosity and $\underline{S}$ is the strain rate tensor.

Now we could write the dissipative term $\underline{\tau} : \vec{\nabla}\vec{V}$ in Eq. (62) by simple transformations:

$$\Phi = \underline{\tau} : \vec{\nabla}\vec{V} = \left(-\frac{2}{3}\mu\left(\vec{\nabla}\cdot\vec{V}\right)\underline{E} + 2\mu\underline{S}\right) : \vec{\nabla}\vec{V} =$$

$$= -\frac{2}{3}\mu\left(\vec{\nabla}\cdot\vec{V}\right)\underline{E} : \vec{\nabla}\vec{V} + 2\mu\cdot\frac{1}{2}\left(\vec{\nabla}\vec{V} + \vec{\nabla}\vec{V}^T\right) : \vec{\nabla}\vec{V} =$$

$$= -\frac{2}{3}\mu\left(\vec{\nabla}\cdot\vec{V}\right)\delta_{ij}\vec{e}_i\vec{e}_j : \frac{\partial V_s}{\partial x_k}\vec{e}_k\vec{e}_s + \mu\left[\vec{\nabla}\vec{V} : \vec{\nabla}\vec{V} + \vec{\nabla}\vec{V}^T : \vec{\nabla}\vec{V}\right] =$$

$$= -\frac{2}{3}\mu\left(\vec{\nabla}\cdot\vec{V}\right)\delta_{ij}\delta_{jk}\delta_{is}\frac{\partial V_s}{\partial x_k} + \mu\left[\frac{\partial V_j}{\partial x_i}\vec{e}_i\vec{e}_j : \frac{\partial V_s}{\partial x_k}\vec{e}_k\vec{e}_s + \frac{\partial V_i}{\partial x_j}\vec{e}_i\vec{e}_j : \frac{\partial V_s}{\partial x_k}\vec{e}_k\vec{e}_s\right] =$$

$$= -\frac{2}{3}\mu\left(\vec{\nabla}\cdot\vec{V}\right)\frac{\partial V_i}{\partial x_i} + \mu\left[\frac{\partial V_j}{\partial x_i}\frac{\partial V_s}{\partial x_k}\delta_{jk}\delta_{is} + \frac{\partial V_i}{\partial x_j}\frac{\partial V_s}{\partial x_k}\delta_{jk}\delta_{is}\right] = -\frac{2}{3}\mu\left(\vec{\nabla}\cdot\vec{V}\right)^2 +$$

$$+\mu\left[\frac{\partial V_j}{\partial x_i}\frac{\partial V_i}{\partial x_j} + \frac{\partial V_i}{\partial x_j}\frac{\partial V_i}{\partial x_j}\right] = -\frac{2}{3}\mu\left(\frac{\partial V_i}{\partial x_i}\right)^2 + \mu\left[\frac{\partial V_j}{\partial x_i}\frac{\partial V_i}{\partial x_j} + \frac{\partial V_i}{\partial x_j}\frac{\partial V_i}{\partial x_j}\right].$$

Thus, in component form Rayleigh function $\Phi$ can be written as

$$\Phi = \underline{\tau} : \vec{\nabla}\vec{V} = -\frac{2}{3}\mu\left(\frac{\partial V_i}{\partial x_i}\right)^2 + \mu\left[\frac{\partial V_j}{\partial x_i}\frac{\partial V_i}{\partial x_j} + \frac{\partial V_i}{\partial x_j} : \frac{\partial V_i}{\partial x_j}\right], \tag{66}$$

or in usual notations

$$\Phi = \mu\left\{-\frac{2}{3}\left(\frac{\partial V_x}{\partial x} + \frac{\partial V_y}{\partial y} + \frac{\partial V_z}{\partial z}\right)^2 + 2\left[\left(\frac{\partial V_x}{\partial x}\right)^2 + \left(\frac{\partial V_y}{\partial y}\right)^2 + \left(\frac{\partial V_z}{\partial z}\right)^2\right]\right.$$
$$\left. + \left(\frac{\partial V_x}{\partial y} + \frac{\partial V_y}{\partial x}\right)^2 + \left(\frac{\partial V_x}{\partial z} + \frac{\partial V_z}{\partial x}\right)^2 + \left(\frac{\partial V_y}{\partial z} + \frac{\partial V_z}{\partial y}\right)^2\right\}. \tag{67}$$

This function can also be written in the componentless form:

$$\Phi = \underline{\tau} : \vec{\nabla}\vec{V} = -\frac{2}{3}\mu\left(\vec{\nabla}\cdot\vec{V}\right)^2 + \mu\left[\vec{\nabla}\vec{V} : \vec{\nabla}\vec{V} + \vec{\nabla}\vec{V} : \vec{\nabla}\vec{V}^T\right]. \tag{68}$$

For perfect gases [6, 7] internal energy is connected with temperature by the relation $du = c_v dT$, where $c_v$ is the isochore thermal capacity. Then, instead of Eq. (62) with the help of expression for vector $\vec{q}$, we can write equation for temperature transport in the form:

$$\rho c_v\left[\frac{\partial T}{\partial t} + \left(\vec{V}\cdot\vec{\nabla}\right)T\right] = \vec{\nabla}\cdot\left(\lambda\vec{\nabla}T\right) - p\left(\vec{V}\cdot\vec{\nabla}\right) + \underline{\tau} : \vec{\nabla}\vec{V} + q_v. \tag{69}$$

The energy equation (Eq. (62)) can be also written in terms of enthalpy $h = u + \frac{p}{\rho}$. With this purpose we need to add the term $\rho\frac{d}{dt}\left(\frac{p}{\rho}\right)$ to the left and right parts of the equation. Then we obtain in the left part $\rho\frac{dh}{dt}$, but in the right part, we shall get this term in transformed form shown as follows:

$$\rho\frac{d}{dt}\left(\frac{p}{\rho}\right) = \frac{dp}{dt} - \frac{p}{\rho}\frac{d\rho}{dt} = \frac{dp}{dt} + p\left(\vec{\nabla}\cdot\vec{V}\right). \tag{70}$$

Here we also used continuity equation (Eq. (2)). Finally, we can obtain energy equation in the form of enthalpy transport as

$$\rho\frac{dh}{dt} = -\vec{\nabla}\cdot\vec{q} + \frac{dp}{dt} + \underline{\tau} : \vec{\nabla}\vec{V} + q_v. \tag{71}$$

This is the second form of energy equation for the perfect gas in which $dh = c_p dT$, where $c_p$ is the isobar heat capacity, which leads to the following transport equation for temperature:

$$\rho c_p\frac{dT}{dt} = \vec{\nabla}\cdot\left(\lambda\vec{\nabla}T\right) + \frac{dp}{dt} + \underline{\tau} : \vec{\nabla}\vec{V} + q_v. \tag{72}$$

One more form of the energy equation can be written if we introduce stagnation enthalpy $h + \frac{V^2}{2}$. To do it we need to add equations for mechanical energy (Eq. (48) to Eq. (71)); as a result we obtain the equation:

$$\rho\frac{d}{dt}\left(h + \frac{V^2}{2}\right) = -\vec{\nabla}\cdot\vec{q} + \frac{\partial p}{\partial t} + \vec{\nabla}\cdot\left(\underline{\tau}\cdot\vec{V}\right) + \rho\vec{V}\cdot\vec{f} + q_v. \tag{73}$$

Here we used the relation:

$$\vec{\nabla}\cdot\left(\underline{\tau}\cdot\vec{V}\right) = \vec{V}\cdot\left(\vec{\nabla}\cdot\underline{\tau}\right) + \underline{\tau} : \vec{\nabla}\vec{V}, \tag{74}$$

which is easy to be proven if we write it down in the component form considering symmetry of stress tensor $\underline{\tau}$.

As a result, the dissipative term in Eq. (73) can be written as follows:

$$\vec{\nabla}\cdot\left(\underline{\tau}\cdot\vec{V}\right) = \Psi + \Phi, \tag{75}$$

where $\Phi$ is the Rayleigh dissipation function and $\Psi = \vec{V}\cdot\left(\vec{\nabla}\cdot\underline{\tau}\right) -$ is the scalar quantity, which can be named as additional dissipation function. This additional dissipation function in Cartesian coordinates for Newtonian Stokes liquid can be written as

$$\Psi = \vec{V}\cdot\left(\vec{\nabla}\cdot\underline{\tau}\right) = \vec{V}\cdot\left\{\vec{\nabla}\cdot\left[-\frac{2}{3}\mu\left(\vec{\nabla}\cdot\vec{V}\right)\underline{E} + 2\mu\underline{S}\right]\right\} =$$

$$= \vec{V}\cdot\left\{\vec{e}_k\frac{\partial}{\partial x_k}\cdot\left[-\frac{2}{3}\mu\left(\vec{\nabla}\cdot\vec{V}\right)\delta_{ij}\vec{e}_i\vec{e}_j + 2\mu S_{ij}\vec{e}_i\vec{e}_j\right]\right\} =$$

$$= \vec{V}\cdot\left\{\frac{\partial}{\partial x_k}\cdot\left[-\frac{2}{3}\mu\left(\vec{\nabla}\cdot\vec{V}\right)\delta_{ij}\delta_{ki}\vec{e}_j + 2\mu S_{ij}\delta_{ki}\vec{e}_j\right]\right\} =$$

$$= \vec{V}\cdot\left\{\frac{\partial}{\partial x_k}\cdot\left[-\frac{2}{3}\mu\left(\vec{\nabla}\cdot\vec{V}\right)\vec{e}_k + 2\mu S_{kj}\vec{e}_j\right]\right\} =$$

$$= V_s\vec{e}_s\cdot\left\{\frac{\partial}{\partial x_k}\left[-\frac{2}{3}\mu\left(\vec{\nabla}\cdot\vec{V}\right)\vec{e}_k + 2\mu S_{kj}\vec{e}_j\right]\right\} =$$

$$= \left(\vec{V}\cdot\vec{\nabla}\right)\left(-\frac{2}{3}\mu\left(\vec{\nabla}\cdot\vec{V}\right)\right) + V_j\frac{\partial}{\partial x_k}\left(2\mu S_{kj}\right).$$

In usual axis designations, the first term is

$$\left(V_x\frac{\partial}{\partial x}+V_y\frac{\partial}{\partial y}+V_z\frac{\partial}{\partial z}\right)\left(-\frac{2}{3}\mu\left(\frac{\partial V_x}{\partial x}+\frac{\partial V_y}{\partial y}+\frac{\partial V_z}{\partial z}\right)\right),$$

while second term is

$$\begin{aligned}V_j\frac{\partial}{\partial x_k}\left(2\mu S_{kj}\right)&=V_1\frac{\partial}{\partial x_k}\left(2\mu S_{k1}\right)+V_2\frac{\partial}{\partial x_k}\left(2\mu S_{k2}\right)+V_3\frac{\partial}{\partial x_k}\left(2\mu S_{k3}\right)=\\&=V_1\left[\frac{\partial}{\partial x_1}\left(2\mu S_{11}\right)+\frac{\partial}{\partial x_2}\left(2\mu S_{21}\right)+\frac{\partial}{\partial x_3}\left(2\mu S_{31}\right)\right]+\\&+V_2\left[\frac{\partial}{\partial x_1}\left(2\mu S_{12}\right)+\frac{\partial}{\partial x_2}\left(2\mu S_{22}\right)+\frac{\partial}{\partial x_3}\left(2\mu S_{32}\right)\right]+\\&+V_3\left[\frac{\partial}{\partial x_1}\left(2\mu S_{13}\right)+\frac{\partial}{\partial x_2}\left(2\mu S_{23}\right)+\frac{\partial}{\partial x_3}\left(2\mu S_{33}\right)\right].\end{aligned}$$

Finally function $\Psi$ in Cartesian coordinates can be written as follows:

$$\begin{aligned}\Psi=&-\frac{2}{3}\left[V_x\frac{\partial}{\partial x}\mu\left(\vec{\nabla}\cdot\vec{V}\right)+V_y\frac{\partial}{\partial y}\mu\left(\vec{\nabla}\cdot\vec{V}\right)+V_z\frac{\partial}{\partial z}\mu\left(\vec{\nabla}\cdot\vec{V}\right)\right]\\&+V_x\left[\frac{\partial}{\partial x}\left(2\mu\frac{\partial V_x}{\partial x}\right)+\frac{\partial}{\partial y}\mu\left(\frac{\partial V_x}{\partial y}+\frac{\partial V_y}{\partial x}\right)+\frac{\partial}{\partial z}\mu\left(\frac{\partial V_z}{\partial x}+\frac{\partial V_x}{\partial z}\right)\right]\\&+V_y\left[\frac{\partial}{\partial x}\mu\left(\frac{\partial V_x}{\partial y}+\frac{\partial V_y}{\partial x}\right)+\frac{\partial}{\partial y}\left(2\mu\frac{\partial V_y}{\partial y}\right)+\frac{\partial}{\partial z}\mu\left(\frac{\partial V_z}{\partial y}+\frac{\partial V_y}{\partial z}\right)\right]\\&+V_z\left[\frac{\partial}{\partial x}\mu\left(\frac{\partial V_z}{\partial x}+\frac{\partial V_x}{\partial z}\right)+\frac{\partial}{\partial y}\mu\left(\frac{\partial V_z}{\partial y}+\frac{\partial V_y}{\partial z}\right)+\frac{\partial}{\partial z}\left(2\mu\frac{\partial V_z}{\partial z}\right)\right].\end{aligned}\tag{76}$$

The fourth form of the energy equation can be written in terms of entropy $s$ transport. According to the fundamental thermodynamic relation,

$$Tds=dh-\frac{1}{\rho}dp.\tag{77}$$

Hence we have

$$T\frac{ds}{dt}=\frac{dh}{dt}-\frac{1}{\rho}\frac{dp}{dt},$$

and then if we substitute quantity $\frac{dh}{dt}$ from Eq. (77) to Eq. (71), we obtain

$$\rho T\frac{ds}{dt}=-\vec{\nabla}\cdot\vec{q}+\underline{\tau}:\vec{\nabla}\vec{V}+q_v.\tag{78}$$

All forms of the equation energy (in terms of internal energy, enthalpy, stagnation enthalpy, and entropy) are equivalent.

Equation for temperature field of an arbitrary gas in the form of equation of transport of temperature $T$ can be obtained from Eq. (62) or Eq. (71). In these cases quantities $\frac{du}{dt}$ and $\frac{dh}{dt}$ for arbitrary gases and liquids must be specified using known formulas, which follow from Maxwell's relations:

$$du = c_v dT - \frac{1}{\rho^2}\left[T\left(\frac{\partial p}{\partial T}\right)_\rho - p\right]d\rho, \tag{79}$$

$$dh = c_p dT + \frac{1}{\rho^2}\left[\rho + T\left(\frac{\partial \rho}{\partial T}\right)_p\right]dp. \tag{80}$$

The subscripts in derivatives here fix the parameters, with the constancy of which the derivatives are calculated. From these formulas the expressions for derivatives can be obtained:

$$\frac{du}{dt} = c_v\frac{dT}{dt} - \frac{1}{\rho^2}\left[T\left(\frac{\partial p}{\partial T}\right)_\rho - p\right]\frac{d\rho}{dt}, \tag{81}$$

$$\frac{dh}{dt} = c_p\frac{dT}{dt} + \frac{1}{\rho^2}\left[\rho + T\left(\frac{\partial \rho}{\partial T}\right)_p\right]\frac{dp}{dt}. \tag{82}$$

If we substitute them into Eq. (62) and Eq. (71), we obtain two forms of the equation energy in terms of temperature transport.

In heat transfer problems, boundary conditions are specified in three different kinds—the first, second, and third kind:

1. The boundary conditions of the first kind consist in setting the temperature on the surface of the body.

2. The boundary conditions of the second kind are setting of the distribution of the heat flux density q on the surface of the body.

3. The boundary conditions of the third kind consist in setting the temperature of the flow over the surface of the body and the heat transfer conditions on its surface.

## 8. Divergence form of transport equations

Material derivative of any physical quantity $\Theta$ multiplied by density $\rho$ always can be written in the "divergent" form as

$$\rho\frac{d\Theta}{dt} = \frac{\partial}{\partial t}\rho\Theta + \vec{\nabla}\cdot\left(\rho\vec{V}\Theta\right). \tag{83}$$

This directly follows from the continuity equation [Eq. (2)].

Let us consider in detail the following cases for three ranks of a certain physical quantity $\Theta$.

### 8.1 Quantity Θ is a scalar

Let us assume that quantity $\Theta$ is temperature *T:*

$$\rho\frac{dT}{dt} = \frac{\partial}{\partial t}\rho T + \vec{\nabla}\cdot\left(\rho\vec{V}T\right). \tag{84}$$

We could prove this equality if we write the left and right parts in component form. For the left part, we have

$$\rho\frac{dT}{dt}=\frac{\partial T}{\partial t}+\rho\left(\vec{V}\cdot\vec{\nabla}\right)T=\rho\frac{\partial T}{\partial t}+\rho V_j\frac{\partial T}{\partial x_j}.$$

For the right part, we have

$$\frac{\partial}{\partial t}\rho T+\vec{\nabla}\cdot\left(\rho\vec{V}T\right)=T\frac{\partial\rho}{\partial t}+\rho\frac{\partial T}{\partial t}+\frac{\partial}{\partial x_j}\rho V_j T=$$

$$=T\frac{\partial\rho}{\partial t}+\rho\frac{\partial T}{\partial t}+TV_j\frac{\partial\rho}{\partial x_j}+\rho T\frac{\partial V_j}{\partial x_j}+\rho V_j\frac{\partial T}{\partial x_j}=$$

$$=T\underbrace{\left(\frac{\partial\rho}{\partial t}+V_j\frac{\partial\rho}{\partial x_j}+\rho\frac{\partial V_j}{\partial x_j}\right)}_{\frac{\partial\rho}{\partial t}+\vec{V}\cdot\vec{\nabla}\rho+\rho\vec{\nabla}\cdot\vec{V}=0}+\rho\frac{\partial T}{\partial t}+\rho V_j\frac{\partial T}{\partial x_j}=\rho\frac{\partial T}{\partial t}+\rho V_j\frac{\partial T}{\partial x_j}.$$

Since the expression in parentheses is zero (due to continuity equation), the equality of the left and right parts is obvious.

### 8.2 Quantity Θ is a vector

Let us assume that quantity Θ is velocity $\vec{V}$. In this case its material derivative can be written in the form:

$$\rho\frac{d\vec{V}}{dt}=\frac{\partial}{\partial t}\rho\vec{V}+\vec{\nabla}\cdot\left(\rho\vec{V}\vec{V}\right). \tag{85}$$

Here in the last term, we see tensor $\rho\vec{V}\otimes\vec{V}$ (momentum flow tensor); the sign of tensor multiplication ⊗ is omitted for ease of recording.

We could prove this equality if we write the left and right parts in the component form and use continuity equation.

For the left part, we have

$$\rho\frac{d\vec{V}}{dt}=\rho\frac{\partial V}{\partial t}+\rho\left(\vec{V}\cdot\vec{\nabla}\right)\vec{V}=\rho\frac{dV_i}{dt}\vec{e}_i+\rho V_j\frac{\partial V_i}{\partial x_j}\vec{e}_i.$$

The first term of the right part is

$$\frac{\partial}{\partial t}\rho\vec{V}=\frac{\partial}{\partial t}\rho V_i\vec{e}_i=\frac{\partial\rho}{\partial t}V_i\vec{e}_i+\rho\frac{\partial V_i}{\partial t}\vec{e}_i.$$

The second term of the right part is

$$\vec{\nabla}\cdot\left(\rho\vec{V}\vec{V}\right)=\vec{e}_k\frac{\partial}{\partial x_k}\cdot\rho V_j V_i\vec{e}_j\vec{e}_i=\frac{\partial}{\partial x_k}\rho V_j V\left(\vec{e}_k\cdot\vec{e}_j\right)\vec{e}_i=$$

$$=\delta_{kj}\frac{\partial}{\partial x_k}\rho V_j V_i\vec{e}_i=\frac{\partial}{\partial x_j}\rho V_j V_i\vec{e}_i=\rho\frac{\partial V_j}{\partial x_j}V_i\vec{e}_i+V_j\frac{\partial\rho}{\partial x_j}V_i\vec{e}_i+\rho V_j\frac{\partial V_i}{\partial x_j}\vec{e}_i.$$

The right part as a whole is

$$\rho\frac{\partial V_i}{\partial t}\vec{e}_i+\rho V_j\frac{\partial V_i}{\partial x_j}\vec{e}_i+V_i\vec{e}_i\underbrace{\left(\frac{\partial\rho}{\partial t}+V_j\frac{\partial\rho}{\partial x_j}+\rho\frac{\partial V_j}{\partial x_j}\right)}_{\frac{\partial\rho}{\partial t}+\vec{V}\cdot\vec{\nabla}\rho+\rho\vec{\nabla}\cdot\vec{V}=0}.$$

Therefore, the expression (Eq. (83)) is valid in case $\Theta$ is a vector.

## 8.3 Quantity Θ is a tensor

Let us assume that $\Theta$ is a tensor, for instance, stress tensor $\underline{\sigma}$. Stress tensor is a second-rank symmetric tensor, which in Cartesian coordinates can be written in the form $\underline{\sigma}=\sigma_{ij}\vec{e}_i\vec{e}_j$. Let us prove the equality:

$$\rho\frac{d\underline{\sigma}}{dt}=\frac{\partial}{\partial t}\rho\underline{\sigma}+\vec{\nabla}\cdot\left(\rho\vec{V}\underline{\sigma}\right). \tag{86}$$

The left part of this equation in the component form can be written as follows:

$$\rho\frac{d\underline{\sigma}}{dt}=\rho\frac{d}{dt}\sigma_{ij}\vec{e}_i\vec{e}_j=\rho\frac{\partial}{\partial t}\sigma_{ij}\vec{e}_i\vec{e}_j+\rho\left(\vec{V}\cdot\vec{\nabla}\right)\sigma_{ij}\vec{e}_i\vec{e}_j=$$
$$=\rho\frac{\partial\sigma_{ij}}{\partial t}\vec{e}_i\vec{e}_j+\rho V_k\frac{\partial\sigma_{ij}}{\partial x_k}\vec{e}_i\vec{e}_j.$$

The right part is

$$\frac{\partial}{\partial t}\rho\underline{\sigma}+\vec{\nabla}\cdot\left(\rho\vec{V}\underline{\sigma}\right)=\frac{\partial}{\partial t}\rho\sigma_{ij}\vec{e}_i\vec{e}_j+\vec{e}_k\frac{\partial}{\partial x_k}\cdot\rho V_s\sigma_{ij}\vec{e}_s\vec{e}_i\vec{e}_j=$$
$$=\sigma_{ij}\frac{\partial\rho}{\partial t}\vec{e}_i\vec{e}_j+\rho\frac{\partial\sigma_{ij}}{\partial t}\vec{e}_i\vec{e}_j+\delta_{ks}\frac{\partial}{\partial x_k}\rho V_s\sigma_{ij}\vec{e}_i\vec{e}_j=$$
$$=\sigma_{ij}\frac{\partial\rho}{\partial t}\vec{e}_i\vec{e}_j+\rho\frac{\partial\sigma_{ij}}{\partial t}\vec{e}_i\vec{e}_j+\frac{\partial}{\partial x_k}\rho V_k\sigma_{ij}\vec{e}_i\vec{e}_j=$$
$$=\sigma_{ij}\frac{\partial\rho}{\partial t}\vec{e}_i\vec{e}_j+\rho\frac{\partial\sigma_{ij}}{\partial t}\vec{e}_i\vec{e}_j+V_k\sigma_{ij}\frac{\partial\rho}{\partial x_k}\vec{e}_i\vec{e}_j+\rho\sigma_{ij}\frac{\partial V_k}{\partial x_k}\vec{e}_i\vec{e}_j+\rho V_k\frac{\partial\sigma_{ij}}{\partial x_k}\vec{e}_i\vec{e}_j=$$
$$=\rho\frac{\partial\sigma_{ij}}{\partial t}\vec{e}_i\vec{e}_j+\rho V_k\frac{\partial\sigma_{ij}}{\partial x_k}\vec{e}_i\vec{e}_j+\sigma_{ij}\vec{e}_i\vec{e}_j\underbrace{\left(\frac{\partial\rho}{\partial t}+V_k\frac{\partial\rho}{\partial x_k}+\rho\frac{\partial V_k}{\partial x_k}\right)}_{=0}$$

Therefore, the expression (Eq. (83)) is also valid in case $\Theta$ is a tensor.

It is necessary to note the derivative $\frac{d}{dt}\underline{\sigma}$, which contains local and convective parts $\frac{d}{dt}=\frac{\partial}{\partial t}+\left(\vec{V}\cdot\vec{\nabla}\right)$, can, at first glance, be used in fluid models when writing the defining equation in the form of a differential transport Equation [2, 5]. However, careful analysis shows that the material derivative for a second-rank tensor is not an invariant quantity [5, 8, 9]. By this reason, instead of derivative $\frac{d}{dt}$, derivative $\frac{D}{Dt}$ is usually used as the material derivative for a second-rank tensor, which contains also rotational part (deviatoric stress rate), which provides symmetry relative to rotations. The rotational part cannot be written in divergent form.

There are different forms of deviatoric stress rate for an arbitrary second-rank tensor $\underline{A}$, for instance:

Jaumann G.: $\underline{A} \cdot \underline{\Omega} + (\underline{A} \cdot \underline{\Omega})^T$, $\underline{\Omega}$—antisymmetric spin tensor.

Rivlin R.: $\underline{A} \cdot \vec{\nabla}\vec{V} + \vec{\nabla}\vec{V}^T \cdot \underline{A}$.

Truesdell C.: $\vec{\nabla}\vec{V} \cdot \underline{A} + \underline{A} \cdot \left(\vec{\nabla}\vec{V}\right)^T + \left(tr\vec{\nabla}\vec{V}\right)\underline{A}$.

Oldroyd J., Sedov L.I., etc. [8, 10, 11]

At present, the question of which derivative is more appropriate to use when constructing rheological equations is unclear. The most common is the rotational derivative by Gustav Jaumann. The corresponding material derivative in the form by Jaumann, for an arbitrary tensor of the second rank, has the form

$$\frac{D\underline{A}}{Dt} = \frac{\partial \underline{A}}{\partial t} + \left(\vec{V} \cdot \vec{\nabla}\right)\underline{A} + \underline{A} \cdot \underline{\Omega} + (\underline{A} \cdot \underline{\Omega})^T. \tag{87}$$

Material derivative in the form by Rivlin is written as follows:

$$\frac{D\underline{A}}{Dt} = \frac{\partial \underline{A}}{\partial t} + \left(\vec{V} \cdot \vec{\nabla}\right)\underline{A} + \underline{A} \cdot \vec{\nabla}\vec{V} + \left(\underline{A} \cdot \vec{\nabla}\vec{V}\right)^T. \tag{88}$$

It is easy to see that Rivlin's derivative differs from Jaumann's one by the additional term $\underline{A} \cdot \underline{S} + (\underline{A} \cdot \underline{S})^T$, which is neutral by itself.

Rotational derivative of a symmetric tensor is also a symmetric tensor. As an example, let us consider the rotational derivative of strain-rate tensor and spin tensor:

$$\underline{S} \cdot \underline{\Omega} + (\underline{S} \cdot \underline{\Omega})^T = \frac{1}{4}\left[\left(\vec{\nabla}\vec{V} + \vec{\nabla}\vec{V}^T\right) \cdot \left(\vec{\nabla}\vec{V} - \vec{\nabla}\vec{V}^T\right) + \left(\vec{\nabla}\vec{V}^T + \vec{\nabla}\vec{V}\right) \cdot \left(\vec{\nabla}\vec{V}^T - \vec{\nabla}\vec{V}\right)\right] =$$
$$= \frac{1}{2}\left[\vec{\nabla}\vec{V}^T \cdot \vec{\nabla}\vec{V} - \vec{\nabla}\vec{V} \cdot \vec{\nabla}\vec{V}^T\right].$$

As a result, we have obtained the symmetrical second-rank tensor.

## 9. Conclusions

In this chapter, some applications of tensor calculus in fluid dynamics and heat transfer are presented. Typical transformations of equations and governing relations are discussed. Main conservation equations are given and analyzed. The governing equations of fluid motion and energy were obtained.

## Author details

Dmitry Nikushchenko* and Valery Pavlovsky
State Marine Technical University, St. Petersburg, Russia

*Address all correspondence to: ndmitry@list.ru

## References

[1] Kochin N, Kibel I, Rose N. Theoretical Hydrodynamic. 6nd Ed. Vol. 1. Moscow: PhysMathLit; 1963. p. 583. (in Russian)

[2] Nikushchenko D, Pavlovsky V. Computational Hydrodynamics. Theoretical Basis. St. Petersburg: Lan; 2018. p. 364. ISBN: 978-5-8114-2924-0 (in Russian)

[3] Batchelor G. An Introduction to Fluid Dynamics. Cambridge: Cambridge University Press; 2000. p. 615

[4] Sedov L. Continuum Mechanics. Moscow: Nauka; 1970. p. 492. (in Russian)

[5] Pavlovsky V. A Short Course in Continuum Mechanics. SPbGTURP: St. Petersburg; 1993. p. 212. (in Russian)

[6] Udaev B. Heat Transfer. Vysshaya shkola: Moscow; 1973. p. 360. (in Russian)

[7] Kutateladze S. Fundamentals of Heat Transfer Theory. Moscow: Atomizdat; 1979. p. 416. (in Russian)

[8] Astarita G, Marrucci G. Principles of Non-Newtonian Fluid Mechanics. Maidenhead–Berkshire: McGraw-Hill; 1974. p. 296

[9] Ehrhard P, editor. Fuhrer Durch Die Stromungslehre: Grundlagen Und Phanomene. Wiesbaden: Springer Fachmedien; 2014. p. 548

[10] Truesdell C, Noll W. The Non-Linear Field Theories of Mechanics. Heidelberg: Springer; 2004. p. 602. ISBN: 978-3-662-10388-3

[11] Oldroyd JG. On the formulation of rheological equations of state. Proceedings of the Royal Society. 1950; **A200**:523

Chapter 5

# Differential Geometry and Macroscopic Descriptions in Nonequilibrium Process

*Claudia B. Ruscitti, Laura B. Langoni and Augusto A. Melgarejo*

## Abstract

The method of Riemannian geometry is fruitful in equilibrium thermodynamics. From the theory of fluctuations it has been possible to construct a metric for the space of thermodynamic equilibrium states. Inspired by these geometric elements, we will discuss the geometric-differential approach of nonequilibrium systems. In particular we will study the geometric aspects from the knowledge of the macroscopic potential associated with the Uhlenbeck-Ornstein (UO) nonequilibrium process. Assuming the geodesic curve as an optimal path and using the affine connection, known as $\alpha$-connection, we will study the conditions under which a diffusive process can be considered optimal. We will also analyze the impact of this behavior on the entropy of the system, relating these results with studies of instabilities in diffusive processes.

**Keywords:** nonequilibrium processes, Uhlenbeck-Ornstein process, statistical manifold, $\alpha$-connections, macroscopic potential

## 1. Introduction

The use of Riemannian geometry associated with the space of thermodynamic equilibrium states has been successful. In this framework the geometric elements are constructed from the knowledge of the thermodynamic potential. In this sense, one of the most interesting ideas is associated with the study of the phase transitions visualized by means of the singularities of the scalar curvature [1, 2]. As a geometric-differential approach of nonequilibrium systems, we consider in our study the geometric properties of a statistical manifold associated with trajectory-dependent entropy [3]. In statistics mechanics it is known that any statistical system has an associated metric-affine manifold having a special affine connection whether it is in equilibrium or not. The affine connection, called the $\alpha$-connection [4], is a generalization of the Levi-Civita connection in Riemannian geometry, and in the case $\alpha = 0$ the metric-affine manifold reduces to the so-called Riemannian manifold. A first element that appears for nonequilibrium systems is the visualization of the phase transitions through the curvature tensor [5]. A second issue is the study of temporal evolution of a statistical system in order to study the optimal evolution.

In other words, it is the analysis of the dynamic behavior of the system and the study of the conditions that are optimal.

In the previous context, this chapter will focus on the analysis of the optimal evolution. In particular we consider the Uhlenbeck-Ornstein (UO) nonequilibrium process described by the probability density function (PDF) solution of the Fokker-Planck Equation [6]. From this probability density function, we build a two-dimensional metric-affine manifold in the coordinates $(\mu, \sigma)$, where $\mu$ is the mean and $\sigma$ the standard deviation. In this coordinates and for the connections $\alpha = 0$ and $\alpha = -1$, the system evolves on a geodesic curve [7].

However, due to simplicity in geometric construction, we are interested in studying the behavior of the system in coordinates $(\theta_1, \theta_2)$, where $\theta_1 = 1/(2\sigma^2)$ and $\theta_2 = -\mu/\sigma^2$. In these coordinates the probability density function belongs to the exponential family, and using this formal expression and analogously with the equilibrium probability density function, we build a macroscopic potential $\psi(\theta_1, \theta_2)$ for the UO process. From the geometry constructed using the $\psi(\theta_1, \theta_2)$ potential, we show that for $\alpha = 3$ and $\alpha = 2$, the system evolves on a geodesic curve. In particular for $\alpha = 3$, we show that the manifold is flat and for the steady state the macroscopic potential and entropy have the same functional dependence. Thinking the geodesic curve as an optimal trajectory, our results allow us to conjecture that the entropy describes the steady state of an optimal evolution.

In the second section of this chapter, we summarize the most relevant aspects of the theory of the statistical manifold. The geometric development associated with the fundamental solution of the Fokker-Planck equation of UO process is found in the third section. The fourth section is devoted to the construction of the potential. In the fifth section, we analyze the geometric relationship between macroscopic potential and entropy. In the sixth section, we present our conclusions and perspectives.

## 2. Elements of statistical manifold

In this section we briefly review the information of geometrical theory [4] that is used to analyze geometrically a family of probability density functions (PDF) and its application to thermodynamics. Let $p(x, \theta)$ be a PDF described by a random variable $x$ and parameters $\theta = (\theta_1, \theta_2, \dots, \theta_m)$ that characterize a system. The set of PDFs

$$M = \{p(x, \theta) : \theta \in \Omega \subset \mathbb{R}^m\} \tag{1}$$

becomes an $m$-dimensional statistical manifold having $\theta_i$ coordinates. According to the information geometrical theory, we can make a metric tensor $g_{ij}$

$$g_{ij}(\theta) = E[\partial_i l(x, \theta) \partial_j l(x, \theta)] = -E[\partial_i \partial_j l(x, \theta)], \tag{2}$$

where $l(x, \theta) = \ln[p(x, \theta)]$ and $E[.]$ means the expectation operation with respect to $p(x, \theta)$. The last expression is obtained by the use of the normalization condition $E[\partial_i l(x, \theta)] = 0$. This metric tensor is the Fisher information matrix in information theory.

In the statistical manifold $M$, we can introduce a natural derivative of the vector field $B$ toward the tangent vector $A$, denoted by $\nabla_A^\alpha B$. It is obtained through the covariant coefficients [7]:

$$\Gamma_{ijk}^{(\alpha)}(\theta) = E\left[\left(\partial_i \partial_j l(x, \theta) + \frac{1-\alpha}{2} \partial_i l(x, \theta) \partial_j l(x, \theta)\right) \partial_k l(x, \theta)\right]. \tag{3}$$

We now restrict our attention to a special family of probability density function, called an exponential family, which is described by [8].

$$p(x,\theta) = \exp\left(C(x) + \sum_i^m \theta_i F_i(x) - \psi(\theta)\right), \tag{4}$$

where $C(x)$ and $F_i(x)$ are arbitrary functions of $x$ and $\psi(\theta)$ is a function of $\theta_i$ coordinates.

Particularly for a probability density function belonging to the exponential family, from Eqs. (2) and (3) the covariant coefficients and metric tensor are written as [9, 10].

$$\Gamma_{ijk}^{(\alpha)}(\theta) = -\frac{(1-\alpha)}{2}\partial_i\partial_j\partial_k\psi(\theta), \tag{5}$$

$$g_{ij}(\theta) = \frac{\partial^2\psi(\theta)}{\partial\theta_i\partial\theta_j}. \tag{6}$$

These coefficients, for $\alpha \in \mathbb{R}$, determine a one-parameter family of affine connections, and each element of this family is called an $\alpha$-connection. An affine connection allows one to compare vectors in nearby tangent spaces [9]. Moreover, the covariant coefficients satisfy the following relation:

$$\Gamma_{ij}^{k(\alpha)} = g^{km}\Gamma_{ijm}^{(\alpha)}, \tag{7}$$

where $\left(g^{km}\right)$ is the inverse matrix of the metric and $\left(g_{ij}\right)$ and $\Gamma_{ij}^{k(\alpha)}$ are the contravariant coefficients.

In the case $\alpha = 0$, the coefficients reduce to the Levi-Civita's connection:

$$\Gamma_{ijk}^{(0)} = \frac{1}{2}\left(\partial_i g_{jk} + \partial_j g_{ik} - \partial_k g_{ij}\right). \tag{8}$$

From Eq. (3) the curvature tensor for an $\alpha$-connection is written as

$$R_{ijkm}^{(\alpha)} = \left(\partial_i\Gamma_{jk}^{s(\alpha)} - \partial_j\Gamma_{ik}^{s(\alpha)}\right)g_{sm} + \left(\Gamma_{irm}^{(\alpha)}\Gamma_{jk}^{r(\alpha)} - \Gamma_{jrm}^{(\alpha)}\Gamma_{ik}^{r(\alpha)}\right). \tag{9}$$

Since $\Gamma_{ijk}^{(\alpha)}(\theta)$ vanishes identically for $\alpha = 1$, any exponential family of PDF constitutes an uncurved space when the $\alpha = 1$ connection is used. The one connection is therefore called the exponential connection as we mentioned above [11]. The relationships (Eqs. (5) and (6)) lead to the simplification of the geometric construction associated with the probability density function (Eq. (4)).

For systems in thermodynamic equilibrium, the parameters $\theta_i$ may include inverse temperature, chemical potential, pressure, magnetic field, and so on. Also, $\psi(\theta)$ represents the thermodynamic potential of the system [8]. It is worth noting that, if we consider the equilibrium density functions and $\alpha = 0$, this formalism reproduces the geometric structure found by Ruppeiner for the space of equilibrium states. In the Ruppeiner formalism, the metric is constructed using the theory of fluctuations, and the geometric elements are obtained as the second derivatives of the corresponding thermodynamic potential [2]. In this paper we have chosen the approach of statistical manifold because it allows us to address nonequilibrium problems.

## 3. Fokker-Planck equation and macroscopic potential

A diffusion process can be thought of as a process of Uhlenbeck-Ornstein. The Uhlenbeck-Ornstein process is a stochastic process that, roughly speaking, describes the velocity of a massive Brownian particle under the influence of friction. The probability density function $P(x,t)$ of the Uhlenbeck-Ornstein process satisfies the Fokker-Planck equation [6]:

$$\frac{\partial P}{\partial t} = \frac{1}{\tau}\frac{\partial}{\partial x}[(x - x_0)P] + D\frac{\partial^2 P}{\partial x^2}. \tag{10}$$

The fundamental solution of this linear parabolic partial differential equation, and the initial condition consisting of a unit point mass at location $y$, is:

$$P(x,t) = \sqrt{\frac{1}{2\pi D\tau(1 - e^{-2t/\tau})}}\exp\left\{-\frac{1}{2D\tau}\left[\frac{\left(x - x_0 - (y - x_0)e^{-t/\tau}\right)^2}{1 - e^{-2t/\tau}}\right]\right\} \tag{11}$$

which is the Gaussian density function with mean

$$\mu = x_0 + (y - x_0)e^{-t/\tau} \tag{12}$$

and variance

$$\sigma^2 = D\tau\left(1 - e^{-2t/\tau}\right) \tag{13}$$

where $x_0$ represents the average length of the displacement, $D$ is the diffusion coefficient, and $\tau$ is a characteristic time. Without loss of generality, in the rest of the work, we consider $y = 0$.

Considering Eqs. (12) and (13), we think the function (Eq. (11)) as a probability density function dependent on two parameters $\mu$ and $\sigma$, formally:

$$p(x,\mu,\sigma) = \frac{1}{\sqrt{2\pi\sigma^2}}e^{-(x-\mu)^2/2\sigma^2}. \tag{14}$$

Considering these parameters as the coordinates $(\mu,\sigma)$ of the manifold $M$ and taking into account Eqs. (2) and (4), it can be seen that for $\alpha = 0$ and $\alpha = -1$, the system evolves on a geodesic curve [7].

Inspired by the simplicity of relations (Eqs. (5) and (6)), we use an alternative description of the UO process through the coordinates:

$$(\theta_1,\theta_2) = \left(\frac{1}{2\sigma^2}, -\frac{\mu}{\sigma^2}\right). \tag{15}$$

In these coordinates the probability density function (Eq. (14)) belongs to the exponential family and is written as

$$p(x,\theta_1,\theta_2) = \exp\left[-\theta_1 x^2 - \theta_2 x - \psi(\theta_1,\theta_2)\right] \tag{16}$$

where

$$\psi(\theta_1,\theta_2) = \frac{\theta_2^2}{4\theta_1} - \ln\left[\sqrt{\frac{\theta_1}{\pi}}\right]. \tag{17}$$

By analogy with the probability density functions for systems in equilibrium, we will call the relationship (Eq. (17)) "nonequilibrium potential" [8].

Using the potential (Eq. (17)) we calculate the coefficients (Eq. (5))

$$\Gamma^{(\alpha)}_{111} = -\frac{(\alpha-1)}{2}\left(\frac{1}{\theta_1^3}+\frac{3\theta_2^2}{2\theta_1^4}\right), \quad \Gamma^{(\alpha)}_{222}=0$$
$$\Gamma^{(\alpha)}_{112} = \Gamma^{(\alpha)}_{121} = \Gamma^{(\alpha)}_{211} = \frac{(\alpha-1)}{2}\frac{\theta_2}{2\theta_1^3} \tag{18}$$
$$\Gamma^{(\alpha)}_{122} = \Gamma^{(\alpha)}_{221} = \Gamma^{(\alpha)}_{212} = -\frac{(\alpha-1)}{4\theta_1^2}$$

Using Eqs. (18) and (19), we calculate the curvature tensor component in the coordinates $(\theta_1, \theta_2)$

$$R^{(\alpha)}_{1212}(\theta_1,\theta_2) = -\frac{(\alpha-3)(\alpha-1)}{8\theta_1^3}. \tag{19}$$

From Eq. (19) we observe that there are two $\alpha$ values for which the manifold is flat, $\alpha = 3$ and $\alpha = 1$. The case $\alpha = 1$ is a direct consequence of Eq. (5).

In a manifold with a connection, we can generalize the straight line of Euclidean geometry. The generalized straight line is called geodesic, and it is defined by the characteristic that its tangent vector does not change its direction. It satisfies the Eq. (11)

$$\frac{d^2\theta_i}{du^2} + \Gamma^i_{jk}\frac{d\theta_k}{du}\frac{d\theta_j}{du} = -\frac{d^2u/ds^2}{(du/ds)^2}\frac{d\theta_i}{du} \tag{20}$$

with an arbitrary parameter $u$. The special parameter $s$ is called the affine parameter. We note that if $u$ is a linear transformation of $s$, the right-hand side vanishes. Let us now prove that the curve defined by Eq. (15) of the UO process is a geodesic. We choose the Newtonian time $t$ for the arbitrary parameter $u$. From Eq. (20) for the study of geodetic curves, we consider that the temporal dependence of the coordinates $(\theta_1, \theta_2)$ is given by

$$\begin{cases} \theta_1(t) = \dfrac{1}{2D\tau(1-e^{-2t/\tau})}, \\ \theta_2(t) = -\dfrac{x_0(1-e^{-t/\tau})}{D\tau(1-e^{-2t/\tau})}. \end{cases} \tag{21}$$

The tangent vector coordinate $d\theta_i/dt$ and the acceleration coordinate $d^2\theta_i/dt^2$ are

$$\frac{d\theta_1}{dt} = -\frac{\cosh^2(t/\tau)}{4D\tau^2}, \quad \frac{d^2\theta_1}{dt^2} = \frac{\coth(t/\tau)\mathrm{csch}^2(t/\tau)}{2D\tau^2} \tag{22}$$

and

$$\frac{d\theta_2}{dt} = -\frac{x_0\mathrm{sech}^2(t/(2\tau))}{4D\tau^2}, \quad \frac{d^2\theta_2}{dt^2} = \frac{2x_0\sinh^4(t/(2\tau))\mathrm{csch}^3(t/\tau)}{D\tau^3}. \tag{23}$$

Substituting Eqs. (22) and (23) in Eq. (20) for the coordinate $\theta_1$, we have

$$\frac{1}{2D\tau^2}\left\{2D\tau[(\alpha-1)-(\alpha-3)\coth(t/\tau)]-(\alpha-1)x_0^2\right\}=\frac{d^2t/ds^2}{(dt/ds)^2}, \tag{24}$$

and for the coordinate $\theta_2$

$$\begin{aligned}&\frac{\operatorname{csch}(t/\tau)}{2D\tau^2}\{2D\tau[-1-(\alpha-2)\cosh(t/\tau)+(\alpha-1)\sinh(t/\tau)]\}+\\&\frac{\operatorname{csch}(t/\tau)}{2D\tau^2}\left[\left(1+e^{-t/\tau}\right)(\alpha-1)x_0^2\right]=\frac{d^2t/ds^2}{(dt/ds)^2}.\end{aligned} \tag{25}$$

The results (Eqs. (24) and (25)) are compatible if we choose $\alpha$ as

$$\alpha=1+2\frac{D\tau}{x_0^2}. \tag{26}$$

Taking as reference the geometric study in the coordinates $(\mu,\sigma)$ [7], we consider two possible relationships between the characteristic time $\tau$ and the other two parameters, $\tau=x_0^2/D$ or $\tau=x_0^2/(2D)$. The first case leads to a flat geometry, that is, if $\tau=x_0^2/D$, we have that $\alpha=3$ and for Eq. (19) $R^{(3)}_{1212}(\theta_1,\theta_2)=0$. This case is equivalent to the connection $\alpha=-1$ in the coordinates $(\mu,\sigma)$, where the manifold is also flat and the affine time $s$ and Newtonian time $t$ are related by the differential equation

$$\frac{1}{\tau}=\frac{d^2t/ds^2}{(dt/ds)^2} \tag{27}$$

whose solution has the form

$$t=-\tau\ln(s+a\tau)+b. \tag{28}$$

On the other hand if $\tau=x_0^2/(2D)$, we have that $\alpha=2$ and the affine and Newtonian time are now related by

$$\frac{\coth(t/\tau)}{\tau}=\frac{d^2t/ds^2}{(dt/ds)^2}. \tag{29}$$

Due to its nature, the equation (Eq. (29)) has been solved numerically. For each $\tau$ we find solutions that make sense in the context of our problem. In **Figure 1** we show two examples of these solutions.

In the coordinates $(\mu,\sigma)$, Eqs. (27) and (29) correspond to the choices $\alpha=-1$ and $\alpha=0$, respectively [7]. On the other hand in the context of geometric construction from the potential (Eq. (17)), the main interpretation of the equation (Eq. (26)) is that the temporal evolution described by Eq. (21) coincides with a geodesic curve of space $(\theta_1,\theta_2)$. In this sense we find two relationships $\tau=x_0^2/D$ and $\tau=x_0^2/(2D)$ between the parameters $x_0$, $D$, and $\tau$ for which the evolution is optimized.

It is interesting to note that if we assume that $v=x_0/\tau$, the relationship $\tau=x_0^2/D$ leads us to $D=v^2\tau$. This is a well-known and a widely used relationship, for example, in the understanding of bacterial mobility through a diffusive model.

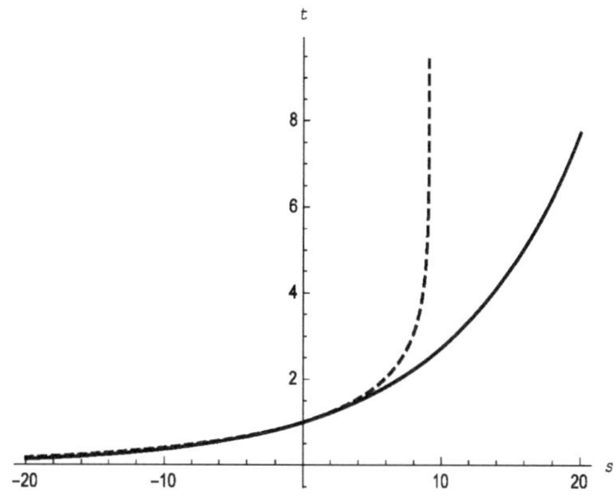


**Figure 1.**
*t as a function of s for two values of τ. τ = 10 in the solid line and τ = 1 in the dashed line.*

In this case the characteristic time $\tau$ represents the tumbling time, and $\nu$ represents the average speed of bacteria [12].

## 4. Optimal trajectory and entropy

Thinking the geodesic curve as an optimal trajectory, from Eq. (21) we will analyze the optimal evolution of the system in terms of the behavior of the macroscopic potential (Eq. (17)) and the entropy function. To begin with this discussion, we first notice that the common definition of a nonequilibrium Gibbs entropy

$$H(\theta) = -\int_{-\infty}^{\infty} p(x,\theta)\ln[p(x,\theta)]dx = \langle h(x,\theta)\rangle \tag{30}$$

suggests to define a trajectory-dependent entropy for the particle (or "system") $h(x,\theta) = -\ln[p(x,\theta)]$ [3]. The trajectory-dependent entropy is related to the function $l(x,\theta)$ used in the construction of the metric of the statistical manifold $M$ by means of the relation $h(x,\theta) = -l(x,\theta)$. In our analysis we are interested in the entropy averaged over trajectories $H(\theta)$.

Moreover, given that the probability density function (Eq. (16)) belongs to the exponential family, we can calculate the entropy by means of the relation [13].

$$H(\theta) = \psi - \theta_1\frac{\partial\psi}{\partial\theta_1} - \theta_2\frac{\partial\psi}{\partial\theta_2}. \tag{31}$$

In any case, using Eqs. (30) or (31), we obtain

$$H(\theta) = \frac{1}{2} - \ln\left[\sqrt{\frac{\theta_1}{\pi}}\right]. \tag{32}$$

In order to study the temporal evolution of potential $\psi$ and entropy $H$, we use the relations Eqs. (12) and (13) in Eq. (21). In particular, we focus on investigating the asymptotic behavior, that is, the behavior for $t \to \infty$. In this limit and using Eq. (21), we have that $\theta_1 = 1/(2D\tau)$ and $\theta_2 = -x_0/(D\tau)$. Therefore, the asymptotic behavior of the potential and entropy is written as

$$\psi = \frac{x_0^2}{2D\tau} + \frac{1}{2}\ln\left[2\pi D\tau\right] \tag{33}$$

and

$$H = \frac{1}{2} + \frac{1}{2}\ln\left[2\pi D\tau\right]. \tag{34}$$

If we think that in the coordinates $(\theta_1, \theta_2)$, with $\alpha = 3$, the system evolves on a geodesic of a flat manifold ($R^{(3)}_{1212}(\theta_1, \theta_2) = 0$), we have $\tau = x_0^2/D$, and both the potential $\psi$ and the entropy $H$ tend to the same asymptotic value. In other words the entropy describes the asymptotic behavior of an optimal evolution:

$$\psi = H = \frac{1}{2} + \ln\sqrt{2\pi D\tau}. \tag{35}$$

In this last result, it can be seen that the asymptotic behavior depends fundamentally on the transport coefficient $D$. In this sense, the relevant variable in the description of the process is the diffusion coefficient $D$. It is interesting to note that the relations (Eq. (21)) are not invariant when we change $t$ by $-t$. In this context, the evolution of the system is irreversible.

While in the coordinates $(\theta_1, \theta_2)$ are two $\alpha$-connections which leads to the system on a geodesic curve ($\alpha = 3$ and $\alpha = 2$), only $\alpha = 3$ makes that the macroscopic potential and entropy have the same functional dependence. It is interesting to note that for this choice of $\alpha$, the manifold in the coordinates $(\theta_1, \theta_2)$ is flat.

In the study of processes, the speed at which they occur is important. In this sense we will study the rate of change of potential:

$$\begin{aligned} \dot{\psi} &= \frac{x_0^2}{4D\tau^2}\operatorname{sech}\left(\frac{t}{2\tau}\right) + \frac{1}{2\tau}\left[\coth\left(\frac{t}{\tau}\right) - 1\right] = \\ &= \frac{1}{2\tau}\operatorname{sech}\left(\frac{t}{2\tau}\right) + \frac{1}{2\tau}\left[\coth\left(\frac{t}{\tau}\right) - 1\right] \geq 0, \end{aligned} \tag{36}$$

and the entropy

$$\dot{H} = \frac{1}{2\tau}\left[\coth\left(\frac{t}{\tau}\right) - 1\right] \geq 0. \tag{37}$$

The equal sign in Eqs. (36) and (37) corresponds to the steady state [14]. In this regard, we associate the asymptotic behavior with the steady state, and we can indicate that the entropy describes the steady state of an optimal evolution.

From the perspective of the probability density function (PDF), when the diffusion coefficient takes small values, the distribution (Eq. (14)) diffuses from a uniform state to a sharp concentrated state, that is, from a stable state to an unstable state [15]. Although in this chapter we have not studied the behavior of the curvature tensor in the $(\mu, \sigma)$ coordinates, these instabilities observed from the perspective of the PDF can be expressed in the singularity of the curvature tensor $R^{(0)}_{1212}(\mu, \sigma)$. An example of this behavior is the formation of density patterns in the populations of self-propelled bacteria whose mobility can be investigated in terms of the diffusion coefficient $D$ [5]. We also observe from Eq. (35) that for $D \to 0$, the entropy and the macroscopic potential have a singular behavior. The behavior that we observe both the tensor $R^{(0)}_{1212}(\mu, \sigma)$ and the potential $\psi$ allows us to conjecture that the instabilities of the system can also be observed in the singularities or discontinuities of the macroscopic potential.

## 5. Discussion and perspectives

In this chapter we have studied the PDF (Eq. (14)) that is the fundamental solution of the Fokker-Planck equation associated with the UO process. Our main interest was relating the geometric aspects of the process with the steady-state behavior. In our analysis we used the theoretical framework of the statistical manifold $M$ with $\alpha$-connections for two different coordinates $(\mu, \sigma)$ and $(\theta_1, \theta_2)$ [4, 7]. In the first case, there exist two interesting values of $\alpha$, namely, $\alpha = -1$ and $\alpha = 0$, for which the process evolves on a geodesic of space $(\mu, \sigma)$ with different values of $\tau$. However, in the search for a simpler geometric construction, we find that for the coordinates $(\theta_1, \theta_2)$, we can define a macroscopic potential $\psi(\theta_1, \theta_2)$, and the values of $\alpha$ that lead to the system evolving on a geodesic curve are $\alpha = 3$ and $\alpha = 2$. In our study we show that the connection $\alpha = -1$ for the coordinates $(\mu, \sigma)$ corresponds to the connection $\alpha = 3$ for the coordinates $(\theta_1, \theta_2)$ in the sense that both connections lead to a flat curvature and the same relationship between the parameters of the system. An important consequence of this behavior is that when the system evolves over geodesics, the macroscopic potential $\psi$ and entropy $H$ have the same functional dependence in the steady state. If we think of the geodesic curve as an optimal trajectory, our results allow us to conjecture that the entropy describes the steady state of an optimal evolution in a flat manifold.

Additionally and poorly developed in this chapter is the use of geometric aspects in the study of instabilities in nonequilibrium system. In equilibrium thermodynamics this information is contained in the scalar curvature of the manifold of equilibrium states [2]. For nonequilibrium problems, the instabilities are associated with the singularities or discontinuities of the curvature tensor [5, 15]. In the case of diffusive problems, the instabilities can be found by studying the singularities of $R^{(0)}_{1212}(\mu, \sigma)$. In terms of diffusivity, we see that as it decreases, the system has different macroscopic behavior. An example of this behavior is the formation of density patterns in the populations of self-propelled bacteria whose mobility can be investigated in terms of the diffusion coefficient $D$ [12]. From the perspective of the macroscopic potential $\psi$, the instabilities can be associated to singularities or discontinuities of $\psi$. In this sense, from a wider point of view, we consider that the potential $\psi$ represents an alternative way to study the phase transitions in nonequilibrium systems.

## Acknowledgements

This work was supported by the Universidad Nacional de La Plata, Argentina (UNLP). AM, LL, and CR are professors at the UNLP.

## Author details

Claudia B. Ruscitti[1†], Laura B. Langoni[2†] and Augusto A. Melgarejo[2*†]

1 Facultad de Ciencias Exactas, Departamento de Matemática, UNLP, La Plata, Argentina

2 Facultad de Ingeniería, Departamento de Ciencias Básicas, UNLP, La Plata, Argentina

*Address all correspondence to: augusto.melgarejo@ing.unlp.edu.ar

† These authors are contributed equally.

## References

[1] Ruppeiner G. Thermodynamics: A Riemannian geometric model. Physical Review A. 1979;**20**:1608-1613

[2] Ruppeiner G. Riemannian geometric approach to critical points: General theory. Physical Review E. 1998;**57**: 5135-5145

[3] Seifert U. Entropy production along a stochastic trajectory and an integral fluctuation theorem. Physical Review Letters. 2005;**95**:040602

[4] Amari S. Differential-geometrical methods in statistics. In: Lecture Notes in Statistics. Vol. 28. Heindelberg: Springer; 1985

[5] Melgarejo A, Langoni L, Ruscitti C. Instability in bacterial populations and the curvature tensor. Physica A: Statistical Mechanics and its Applications. 2016;**458**:189-193

[6] Uhlenbeck G, Ornstein L. On the theory of the Brownian motion. Physics Review. 1930;**36**:823-841

[7] Ruscitti C, Langoni L, Melgarejo A. Description of instabilities in Uhlenbeck–Ornstein process. Physica Scripta. 2019;**94**:115010

[8] Brody D, Hook D. Information geometry in vapour-liquid equilibrium. Journal of Physics A: Mathematical and Theoretical. 2009;**42**:023001

[9] Amari S. Differential geometry of curved exponential families-curvatures and information loss. Annals of Statistics. 1982;**10**:357-385

[10] Oshima H, Obata T, Hara H. Riemann scalar curvature of ideal quantum gases obeying Gentile's statistics. Journal of Physics A: Mathematical and General. 1999;**32**: 6373-6383

[11] Zhang J. On monotone embedding in information geometry. Entropy. 2015; **17**:4485-4499

[12] Cates M, Marenduzzo D, Pagonabarraga I, Tailleur J. Arrested phase separation in reproducing bacteria creates a generic route to pattern formation. Proceedings of the National Academy of Sciences. 2010; **107**:11715-11720

[13] Jaynes ET. Information theory and statistical mechanics. Physics Review. 1957;**106**:620-630

[14] Kondepudi D, Prigogine I. Modern Thermodynamics from Heat Engines to Dissipative Structures. 2nd ed. United Kingdom: John Wiley & Sons; 2015

[15] Obata T, Hara H, Endo K. Differential geometry of non-equilibrium processes. Physical Review A. 1992;**45**:6997-7001

## Section 4

# Advanced Topics of Tensor Analysis

Chapter 6

# Derived Tensor Products and Their Applications

*Francisco Bulnes*

## Abstract

In this research we studied the tensor product on derived categories of Étale sheaves with transfers considering as fundamental, the tensor product of categories $X \otimes Y = X \times Y$, on the category $\mathrm{Cor}_k$, (finite correspondences category) by understanding it to be the product of the underlying schemes on $k$. Although, to this is required to build a total tensor product on the category $\mathrm{PST}(k)$, where this construction will be useful to obtain generalizations on derived categories using pre-sheaves and contravariant and covariant functors on additive categories to define the exactness of infinite sequences and resolution of spectral sequences. Some concrete applications are given through a result on field equations solution.

**Keywords:** algebraic variety, additive pre-sheaves, derived categories, derived tensor products, finite correspondences category, schemes, singularities, varieties
**2010 Mathematics Subject classification**: 13D09, 18D20, 13D15

## 1. Introduction

This study is focused on the derived tensor product whose functors have images as cohomology groups that are representations of integrals of sheaves represented for its pre-sheaves in an order modulo $k$. This study is remounted to the $K$-theory on the sheaves cohomologies constructed through pre-sheaves defined by the tensor product on commutative rings. The intention of this study is to establish a methodology through commutative rings and their construction of a total tensor product $\otimes^L$, [1]on the category $\mathrm{PST}(k)$, considering extensions of the tensor products $\otimes_{R(\mathcal{A})}$, to obtain resolution in the projective sense of infinite sequences of modules of Étale sheaves. These sheaves are pre-sheaves of Abelian groups on the category of smooth separated schemes restricted to scheme X.

Likewise, the immediate application of the derived tensor products will be the determination of the tensor triangulated category $\mathrm{DM}^-_{ett}(k, \mathbb{Z}/m)$, of Étale motives to be equivalent to the derived category of discrete $\mathbb{Z}/m$- modules over the Galois

[1] $\mathbb{L}$, is a Lefschetz motive $\mathbb{Z}(1)$, [1].

group $G = \mathrm{Gal}(k_{sep}/k)$, which says on the equivalence of functors of tensor triangulated categories[2].

Then the mean result of derived tensor products will be in tensor triangulated category $\mathrm{DM}_{Nis}^{\mathrm{eff},-}(k, R)$, of effective motives and their subcategory of effective geometric motives $\mathrm{DM}_{gm}^{\mathrm{eff},-}(k, R)$. Likewise, the motive $M(\mathrm{X})$, of a scheme $X$, is an object of $\mathrm{DM}_{Nis}^{\mathrm{eff},-}(k, R)$, and belongs to $\mathrm{DM}_{gm}^{\mathrm{eff},-}(k, R)$, if $X$, is smooth. However, this requires the use of cohomological properties of sheaves associated with homotopy invariant pre-sheaves with transfers for Zariski topology, Nisnevich and cdh topologies.

Finally, all this treatment goes in-walked to develop a motivic cohomology to establish a resolution in the field theory incorporating singularities in the complex Riemannian manifolds where singularities can be studied with deformation theory through operads, motives, and deformation quantization.

## 2. Fundaments of derived tensor products

We consider the Abelian category Ab, which is conformed by all functor images that are contravariant additive functors $F : \mathcal{A} \to \mathrm{Ab}$, on small category of $\mathbb{Z}(\mathcal{A})$. Likewise, $\mathbb{Z}(\mathcal{A})$, is the category of all additive pre-sheaves on $\mathcal{A}$. Likewise, we can define this category as of points space:

$$\mathbb{Z}(\mathcal{A}) = \{F | F : \mathcal{A} \to \mathrm{Ab}\}, \tag{1}$$

Likewise, we have the Yoneda embedding as the mapping[3]:

$$h : \mathcal{A} \to \mathbb{Z}(\mathcal{A}), \tag{2}$$

which has correspondence rule

$$\mathrm{X} \mapsto \oplus \mathrm{X}_i, \tag{3}$$

or

$$h_{\mathrm{X}} = \oplus h_{\mathrm{X}_i}, \tag{4}$$

We need a generalization of the before categories and functors, therefore we give a ring $R$, originating the ring structure $\mathcal{A}(R)$, to be the Abelian category of the additive functors

$$F : \mathcal{A} \to R - \mathrm{mod}, \tag{5}$$

being $R$-mod, the category of the modules that originate the ring structure. Then $h_X$, is the functor

---

2 **Theorem.** If $1/\mathrm{m} \in k$, the space $(\mathcal{L}, \otimes_{\mathcal{L}})$, is a tensor triangulated category and the functors

$$\mathrm{D}^{-}(G, \mathbb{Z}/m) \overset{\pi *}{\to} \mathcal{L} \to \mathrm{D}(W_A^{-1}) = \mathrm{DM}_{Et}^{\mathrm{eff},-}(k, \mathbb{Z}/m),$$

until the category $\mathrm{D}^{-}(Sh_{et}(Cor_k, \mathbb{Z}/m))$.

3 The obtained image by the Yoneda embedding has the pre-sheaf $\mathcal{A}^{\oplus} \subset \mathbb{Z}(\mathcal{A})$.

$$h_X : A \mapsto R \otimes_{\mathbb{Z}} \mathrm{Hom}_{\mathcal{A}}(A, X), \tag{6}$$

which is representable of the $R$-mod.

Likewise, the following lemma introduces the representable pre-sheaves and functors and their role to construct pre-sheaves $\otimes_{R(\mathcal{A})}$, that can be extended to pre-sheaves $\otimes^L$, first using the projective objects of $R(\mathcal{A})$, and define the projective resolution to infinite complexes sequence.

**Lemma 1.1.** Every representable pre-sheaf $h_X$, is a projective object of $R(\mathcal{A})$, every projective object of $R(\mathcal{A})$ is a direct summand of a direct sum of representable functors, and every $F$, in $R(\mathcal{A})$, has a projective resolution.

*Proof*. We consider an analogue to (6) in the functor context:

$$\mathrm{Hom}_{R(\mathcal{A})}(h_X, F) \cong F(X), \tag{7}$$

Then each object $h_X$, is a projective object in $R(\mathcal{A})$. Likewise, each $F \in R(\mathcal{A})$, is a quotient

$$F = h_X / \mathcal{A}^{\oplus}, \tag{8}$$

then there exist a surjection x, such that

$$\mathrm{x} : \oplus h_X \rightarrow F, \tag{9}$$

Then from the additive category until functional additive category modulus $\mathcal{A}^{\oplus} \subset \mathbb{Z}(\mathcal{A})$, we have:

$$F = \bigoplus_{X \in \mathcal{A}} \bigoplus_{\substack{\mathrm{x} \in F(X) \\ \mathrm{x} \neq 0}} h_X, \tag{10}$$

which proves the lemma. ■

Now suppose that $\mathcal{A}$, with an additive symmetric monoidal structure $\otimes$, is such that

$$\mathcal{A} = Cor_k, \tag{11}$$

This means that $\otimes$, commutes with direct sum. Let $N_\alpha$, $\alpha \in A$, and $M$, be $R$-modules; then is clear that:

$$M \otimes \left( \bigoplus_{\alpha \in A} N_\alpha \right) \cong \bigoplus_{\alpha \in A} (M \otimes N_\alpha), \tag{12}$$

We extend $\otimes$, on $\mathcal{A}^{\oplus}$, in the same way, and this extends to tensor product of corresponding projectives. Then $\otimes$, can be extended to a tensor product on all of $R(\mathcal{A})$.

Likewise, if $F, G \in R(\mathcal{A})$, then we have a pre-sheaf tensor product in the following way:

$$(F \otimes_R G)(X) = F(X) \otimes_R G(X), \tag{13}$$

However, this does not correspond to $R(\mathcal{A})$, since $F \otimes_R G$, is not additive. However, this could be additive when one component $F$, or $G$, is element of $\mathcal{A}^{\oplus}$. But if we want to get a tensor product on $R(\mathcal{A})$, we need a more complicated or specialized construction. For this, we consider $X, Y \in \mathcal{A}$, then $h_X \otimes h_Y$, of their representable pre-sheaves should be represented by $X \otimes Y$. As a first step, we can extend $\otimes$, to a tensor product

$$\otimes : \mathcal{A}^{\oplus} \times \mathcal{A}^{\oplus} \to \mathcal{A}^{\oplus}, \tag{14}$$

commuting with $\oplus$. Thus if $L_1, L_2 \in \mathrm{Ch}^{-}(\mathcal{A}^{\oplus})$, of the above co-chain complexes as follows:

$$\ldots \to F^n \to 0 \to \ldots, \tag{15}$$

the chain complex $L_1 \otimes L_2$, is defined as the total complex of the double complex $L_1^* \otimes L_2^*$.

Then we can define a legitimate tensor product between two categories $F, G \in R(\mathcal{A})$, as follows:

**Definition 1.1.** Let be $F, G \in R(\mathcal{A})$, choosing projective resolutions

$$P_* \to F, \quad Q_* \to G, \tag{16}$$

we define $F \otimes^L G$, [4]to be $P \otimes Q$, which means that the tensor product is total having that $\mathrm{Tot}(P_* \otimes Q_*)$. Then the tensor product to these pre-sheaves and the Hom, pre-sheaves is defined as:

$$F \otimes G = H_0\left(F \otimes^L G\right), \tag{17}$$

and

$$\mathrm{Hom}(F, G) : X \mapsto \mathrm{Hom}_{R(\mathcal{A})}(F \otimes h_X, G), \tag{18}$$

The relation (17) means the chain homotopy equivalent of the $F \otimes^L G$, is well defined up to chain homotopy equivalence, and analogous for $\mathrm{Hom}(F, G)$.

In particular, given that $h_X$, and $h_Y$, are projective, we have

$$h_X \otimes^L h_Y = h_X \otimes h_Y = h_{X \otimes Y}, \quad \forall X, Y \in \mathcal{A}^{\oplus}. \tag{19}$$

Likewise, the ring $R(\mathcal{A})$, is an additive symmetric monoidal category.

We consider the following lemma.

Lemma 1.2. The functor $\mathrm{Hom}(F, \bullet)$, is right adjoint to $F \otimes \bullet$. In particular $\mathrm{Hom}(F, \bullet)$, is left exact and $F \otimes \bullet$, is right exact.

*Proof*. Let be

$$F = \mathrm{Hom}(h_Y, G), \tag{20}$$

Then[5]

$$\mathrm{Hom}_{R(\mathcal{A})}(h_X, F) = \mathrm{Hom}_{R(\mathcal{A})}(h_X, \mathrm{Hom}(h_Y, G)), \tag{21}$$

We consider

$$\mathrm{Hom}(h_Y, G) = \mathrm{Hom}_{R(\mathcal{A})}(h_X \otimes h_Y, G), \tag{22}$$

[4] $\otimes^L$, is a total tensor product.

[5] $\mathrm{Hom}(h_Y, G) : Y \to \mathrm{Hom}_{R(\mathcal{A})}(F \otimes h_Y, G)$.

Then in (20) we have:

$$\mathrm{Hom}_{R(\mathcal{A})}\left(h_{\mathrm{X}}, \mathrm{Hom}_{R(\mathcal{A})}(h_{\mathrm{X}} \otimes h_{\mathrm{Y}}, G)\right) = G(\mathrm{X} \otimes \mathrm{Y}), \tag{23}$$

But also,

$$G(\mathrm{X} \otimes \mathrm{Y}) = \mathrm{Hom}_{R(\mathcal{A})}(h_{\mathrm{X}} \otimes h_{\mathrm{Y}}, \mathrm{G}), \tag{24}$$

where the lemma is proved. ■

We consider the following examples.

**Example 1.1.** We consider the category $\mathcal{A}$, of free $R$-modules over a commutative ring $R(\mathcal{A})$. This category is equivalent to the category of all $R$-modules where pre-sheaf associated to $M$, is $M \otimes_R$, and Hom, and $\otimes$, are the familiar $\mathrm{Hom}_R$, and $\otimes_R$.

Here, for any two modules $A, B \in R(\mathcal{A})$, we have:

$$A \otimes_R B : A \otimes B = B \otimes A, \tag{25}$$

**Example 2.1.** Let $\mathcal{A}$, be the category of $R$-modules $M$, such that:

$$K \otimes_R M \cong K \otimes_R (M/M_{\mathrm{tor}}), \tag{26}$$

where $K$, is a fraction field[6] and $M_{\mathrm{tor}}$, is the torsion submodule of $M$. Then associated to $M$, is $1 \otimes_R M$, which is pre-sheaf. Here Hom. and $\otimes$, are $\mathrm{Hom}_R$, and $\otimes_{\mathrm{tor}}$.

**Example 3.1.** Let $R$, be a simplicial commutative ring and $R(\mathcal{A}) \to \mathcal{A}$, be a category cofibrant replacement. Here, the pre-sheaf associated to $M$, which is the Kähler 1-differentials module, is $M \otimes^{\mathrm{L}}_{R(\mathcal{A})}$, and here Hom, and $\otimes$, are $\mathrm{Hom}_R$, and $\otimes^{L}_{R(\mathcal{A})}$. Here the category is of the cotangent complexes of $R$.

**Proposition 1.1.** If $F_i$, and $G_i$, are in $R(\mathcal{A})$, then there is a natural mapping

$$\mathrm{Hom}(F_1, G_1) \otimes \mathrm{Hom}(F_2, G_2) \to \mathrm{Hom}(F_1 \otimes G_1, F_2 \otimes G_2), \tag{27}$$

compatible with the monoidal pairing

$$\begin{aligned}\mathrm{Hom}_{\mathcal{A}}(U \times A_1, \mathrm{X}_1) \otimes \mathrm{Hom}_{\mathcal{A}}(U \times A_2, \mathrm{X}_2) &\to \mathrm{Hom}_{\mathcal{A}}(U \times U \times A_1 \times A_2, \mathrm{X}_1 \times \mathrm{X}_2)\\ &\to \mathrm{Hom}_{\mathcal{A}}(U \times A_1 \times A_2, \mathrm{X}_1 \times \mathrm{X}_2),\end{aligned} \tag{28}$$

*Proof*. We have Hom, as defined in (18):

$$\mathrm{Hom}(F_1, G_1) : \mathrm{X}_1 \to \mathrm{Hom}_{R(\mathcal{A})}(F_1 \otimes h_{\mathrm{X}_1}, G_1), \tag{29}$$

and

$$\mathrm{Hom}(F_2, G_2) : \mathrm{X}_2 \to \mathrm{Hom}_{R(\mathcal{A})}(F_2 \otimes h_{\mathrm{X}_2}, G_2), \tag{30}$$

If $F_1, G_1, F_2, G_2 \in R(\mathcal{A})$., then

[6] The field of fractions of an integral domain is the smallest field in which this domain can be embedded.

$$\begin{aligned}\mathrm{Hom}(F_1,G_1)\otimes\mathrm{Hom}(F_2,G_2) &= \mathrm{Hom}_{R(\mathcal{A})}(F_1\otimes h_{X_1}\otimes F_2\otimes h_{X_2},G_1\otimes G_2)\\ &= \mathrm{Hom}_{R(\mathcal{A})}(F_1\otimes F_2\otimes h_{X\otimes Y},G_1\otimes G_2)\\ &= \mathrm{Hom}(F_1\otimes F_2,G_1\otimes G_2),\end{aligned} \tag{31}$$

We consider the Universal mapping which is commutative:

$$\begin{array}{ccc} F\times G & \xrightarrow{\lambda} & F\otimes G \\ & \mathrm{f}\searrow & \\ & & S \end{array}$$

Then (31) is compatible with the monoidal pairing. ■

If the (projective) objects $h_X$, are flat, that is to say, $h_X\otimes\bullet$, is an exact functor then $\otimes$, is called a balanced functor [2]. Here $F\otimes^L G$, agrees with the usual left derived functor $L(F\otimes\bullet)G$. But here we do not know when the $h_X$, are flat. This is true in Example 1.1. But it is not true in $\mathrm{PST}(k)=\mathbb{Z}(Cor_k)$. Then we need to extend $\otimes^L$, to a total tensor product on the category $\mathrm{Ch}^-R(\mathcal{A})$, of bounded above co-chain complexes (15). This would be the usual derived functor if $\otimes$, were balanced [2], and our construction is parallel. Likewise, if $C$, is a complex in $\mathrm{Ch}^-R(\mathcal{A})$, there is a quasi-isomorphism $P\cong\rightarrow C$, with $P$, a complex of projective objects. Any such complex $P$, is called a projective resolution of $C$, and any other projective resolution of $C$, is chain homotopic to $P$ [3].

Likewise, if $D$, is any complex in $\mathrm{Ch}^-R(\mathcal{A})$, and

$$Q\overset{\cong}{\rightarrow}D, \tag{32}$$

is a projective resolution, we define

$$C\otimes^L D=P\otimes Q, \tag{33}$$

Now, how do we understand the extensions of these tensor products in chain homotopy equivalence?

Since $P$, and $Q$, are bounded above, each

$$(P\otimes Q)^n=\bigoplus_{i+j=n}P^i\otimes Q^j, \tag{34}$$

is a finite sum, and $C\otimes^L D$, is bounded above. Then, since $P$, and $Q$, are defined up to chain homotopy, the complex $C\otimes^L D$, is independent (up to chain homotopy equivalence) of the choice of $P$, and $Q$. Then there exists a mapping

$$C\otimes^L D\rightarrow C\otimes D, \tag{35}$$

which extends the mapping

$$F\otimes^L G\rightarrow F\otimes G, \tag{36}$$

of Definition 1.1.

We consider the following lemma to obtain in the extension (36) a derived triangulated category that will be useful in the context of derived tensor categories whose pre-sheaves are Étale pre-sheaves.

The importance of a triangulated category together with the additional structure as the given by pre-sheaves $\otimes^L$, lies in obtaining distinguished triangles of categories that generate the long exact sequences of homology that can be described through of short exact sequences of Abelian categories. Likewise, the immediate examples are the derived categories of Abelian category and the stable homotopy category of spectra or more generally, the homotopy category of a stable $\infty$-category. In both cases is carried a structure of triangulated category.

## 3. Derived triangulated categories with structure by pre-sheaves $\otimes^L$, and $\otimes^{tr}_{L,ét}$

We enounce the following proposition.

**Proposition 3.1.** The derived category $\mathrm{D}^-R(\mathcal{A})$, equipped with $\otimes^L$- structure is a tensor-triangulated category.

*Proof*. We consider a projective object $\mathrm{X}\in\mathcal{P}$, where $\mathcal{P}$, is a projective category defined as the points set

$$\mathcal{P}=\{\mathrm{X}\in R(\mathcal{A})|\mathcal{A}\text{ is addative with }\otimes_R-\text{structure}\}, \tag{37}$$

We consider the application $\Lambda$, defined by the mapping:

$$\Lambda:\mathcal{P}\to\mathrm{K}^-(\mathcal{P}), \tag{38}$$

where the objects $\Lambda(\mathrm{X})$, are those that are determined by

$$H_0\left(\Lambda\otimes^L Q\right)=\mathrm{Hom}(h_\mathrm{X},\mathrm{Hom}(\Lambda\otimes h_\mathrm{Y},G)), \tag{39}$$

or

$$\Lambda(\mathrm{X})=\mathrm{Hom}_{R(\mathcal{A})}(h_\mathrm{X},\Lambda)\in\mathrm{D}^-R(\mathcal{A}), \tag{40}$$

Then we have

$$\mathrm{D}^-R(\mathcal{A})\cong\mathrm{K}^-(\mathcal{P}), \tag{41}$$

via the chain homotopy. For other side

$$\mathrm{X}\to\mathrm{Hom}(h_\mathrm{X},\mathrm{Hom}(\Lambda\otimes h_\mathrm{Y},G)), \tag{42}$$

which is risked from $\otimes^L-$ structure when $\otimes\cong\otimes^L$, in $\mathcal{P}$, which then is true from the lemma 2.1. ■

Now, for bounded complexes of pre-sheaves we can give the following definitions.

**Definition 3.1**. Let $C$, and $D$, be bounded complexes of pre-sheaves. There is a canonical mapping:

$$C\otimes_R D\to C\otimes D, \tag{43}$$

which was foresee in the Definition 1.1. By right exactness of $\otimes_R$, and $\otimes$, given in Lemma 1.1, it suffices to construct a natural mapping of pre-sheaves

$$\eta : h_X \otimes_R h_Y \rightarrow h_{X\otimes Y},$$
$$\forall X, Y \in \mathcal{A}^{\oplus}. \tag{44}$$

For $U$, in $\mathcal{A}$, $\eta_U$, is the monoidal product in $\mathcal{A}$, followed by the diagonal mapping of triangle:

$$\eta : U \rightarrow U \otimes U, \tag{45}$$

that is to say,

$$h_X(U) \otimes_R h_Y(U) \rightarrow \mathrm{Hom}_{\mathcal{A}}(U, X) \otimes_R \mathrm{Hom}_{\mathcal{A}}(U, Y), \tag{46}$$

satisfies the triangle[7]:

$$\begin{array}{ccc} \mathrm{Hom}_{\mathcal{A}}(U, X) \otimes_R \mathrm{Hom}_{\mathcal{A}}(U, Y) & \overset{\otimes}{\rightarrow} & \mathrm{Hom}_{\mathcal{A}}(U \otimes U, X \otimes Y) \\ \Delta\searrow & & \downarrow\Delta \\ & & \mathrm{Hom}_{\mathcal{A}}(U, X \otimes Y) \end{array} \tag{47}$$

where

$$\mathrm{Hom}_{\mathcal{A}}(U, X \otimes Y) = h_{X\otimes Y}(U), \tag{48}$$

With all these dispositions and generalities, now we can specialize to the case when[8]

$$\mathcal{A} = Cor_k, \tag{49}$$

and $\otimes$, is the tensor product

$$X \otimes Y = X \times Y', \tag{50}$$

Then we have the Yoneda embedding:

$$\mathrm{PST}(k) \subset Cor_k{}^{\oplus} \subset Cor_k, \tag{51}$$

We denote as $\otimes^{tr}$, for the tensor product on $\mathrm{PST}(k) = \mathbb{Z}(Cor_k)$, or

$$\mathrm{PST}(k, R) = R(Cor_k), \tag{52}$$

and $\otimes^{tr}{}_L$, for $\otimes^L$. Then there is a natural mapping

$$C \otimes_L^{tr} D \rightarrow C \otimes^{tr} D, \tag{53}$$

Here $\otimes^{tr}{}_L$, is the tensor product induced to $\mathbb{Z}(Cor_k)$. But, before we will keep using the product $\otimes^{tr}$, which we can define as:

$$X \otimes Y = X \times Y.$$
$$R_{tr}(X) \otimes^{tr} R_{tr}(Y) = R_{tr}(X, Y), \tag{54}$$

being $h_X = R_{\mathrm{tr}}(X)$, $\forall h_X \in \mathrm{Hom}$, $\forall X \in \mathcal{A}^{\oplus}$.

---

[7] $\eta_U \circ \Delta = \Delta'$.

[8] Def. If X, $Y \in Cor_k$, their tensor product $X \otimes Y$, is defined to be the product underlying schemes over $k$,

$$X \otimes Y = X \times Y.$$

The above can be generalized through the following lemma.

**Lemma 3.1.** The pre-sheaf $\mathbb{Z}_{tr}((X_1, x_1)\wedge\cdots\wedge(X_n, x_n))$, is a direct summand of $\mathbb{Z}_{tr}(X_1 \times \cdots \times X_n)$. In particular, it is projective object of PST. Likewise, for the following sequence of pre-sheaves with transfers, the exactness is explicit[9]:

$$\begin{aligned}
0 &\to \mathbb{Z}_{tr}(X_1 \times \cdots \times X_n) \to \bigoplus_i \mathbb{Z}_{tr}(X_i) \to \bigoplus_{i,j} \mathbb{Z}_{tr}(X_i \times X_j) \to \dots \\
&\to \bigoplus_{i,j} \mathbb{Z}_{tr}(X_1 \times \dots \times X_j \times \dots \times X_j \times \dots \times X_n) \to \bigoplus_i \mathbb{Z}_{tr}(X_1 \times \dots \times \hat{X}_j \times \dots \times X_n) \\
&\to \mathbb{Z}_{tr}(X_1 \times \cdots \times X_n) \to \mathbb{Z}_{tr}(X_1\wedge\cdots\wedge X_n) \to 0,
\end{aligned} \tag{55}$$

Then, it is sufficient to demonstrate that $\otimes^{tr}_{L,ét}$, preserve quasi-isomorphisms.

**Definition 3.2.** A pre-sheaf with transfers is a contravariant additive functor:

$$F = Cor_k \to \text{Ab}, \tag{56}$$

we write

$$\text{Pre}Sh(Cor_k) \to \text{PST}(k) = \text{PST}, \tag{57}$$

to describe the functor category on the field $k$, whose objects are pre-sheaves with transfer and whose morphisms are natural transformations.

Likewise, analogously we can define to the tensor product $\otimes^{tr}$, their extension to $\otimes^{tr}{}_{ét}$.

Likewise, we have the definition.

**Definition 3.3.** If $F$, and $G$, are pre-sheaves of $R$-modules with transfers, we write:

$$(F \otimes^{tr} G)_{ét} \to F \otimes^{tr}_{ét} G, \tag{58}$$

the Étale sheaf associated to $F \otimes^{tr} G$.

If $C$, and $D$, are bounded above complexes of pre-sheaves with transfers, we shall write $C \otimes^{tr}_{ét} D$, for $(C \otimes^{tr} D)_{ét}$, and

$$\left(C \otimes^{tr}_{L} D\right) \cong P \otimes^{tr}_{ét} Q, \tag{59}$$

where $P$, and $Q$, are complexes of representable sheaves with transfers, $P \cong C$, and $Q \cong D$. Then there is a natural mapping

$$\left(C \otimes^{tr}_{L,ét} D\right) \to C \otimes^{tr}_{ét} D, \tag{60}$$

induced by

$$\left(C \otimes^{tr}_{L,} D\right) \to C \otimes^{tr} D, \tag{61}$$

**Lemma 3.2.** If $F$, and $F'$, are Étale sheaves of $R$-modules with transfers, and $F$, is locally constant, the mapping:

$$\begin{aligned}
h_X(U) \otimes_R h_Y(U) = \text{Hom}_{\mathcal{A}}(U, X) \otimes \text{Hom}_{\mathcal{A}}(U, Y) &\overset{\otimes}{\to} \text{Hom}_{\mathcal{A}}(U \otimes U, X \otimes Y) \\
&\overset{\Delta'}{\to} \text{Hom}_{\mathcal{A}}(U, X \otimes Y) = h_{X \otimes Y}(U),
\end{aligned} \tag{62}$$

---

9 $\mathbb{Z}_{tr}(X) \cong \mathbb{Z}_{tr} \oplus \mathbb{Z}_{tr}(X, x)$, $\mathbb{Z}_{tr}(X_1 \times X_2) \cong \mathbb{Z}_{tr} \oplus \mathbb{Z}_{tr}(X_1, x_1) \oplus \mathbb{Z}_{tr}(X_2, x_2) \oplus \mathbb{Z}_{tr}(X_1 \wedge X_2)$.

induces an isomorphism

$$F \otimes_{\mathrm{et}} F' \overset{\cong}{\rightarrow} F \otimes_{\acute{e}t}^{tr} F', \tag{63}$$

Remember that a pre-sheaf is defined as:

**Definition 3.4.** A pre-sheaf $F$, of Abelian groups on $Sm/k$, is an Étale sheaf if it restricts to an Étale sheaf on each X, in $Sm/k$, , that is if:

i. The sequence

$$0 \rightarrow F(\mathrm{X}) \overset{\mathrm{diag}}{\rightarrow} F(U) \overset{(+,-)}{\rightarrow} F(U \times_{\mathrm{X}} U), \tag{64}$$

is exact for every surjective Étale morphism of smooth schemes,

$$U \rightarrow \mathrm{X}, \tag{65}$$

ii. $F(\mathrm{X} \underline{\cup} \mathrm{Y}) = F(\mathrm{X}) \oplus F(\mathrm{Y}), \ \forall \mathrm{X}, \mathrm{Y},$ schemes.

We demonstrate Lemma 3.2.

*Proof*. We want the tensor product $\otimes_{L,\acute{e}t}^{tr}$, which induces to tensor triangulated structure on the derived category of Étale sheaves of $R$-modules with transfers[10] defined in other expositions [4]. Considering Proposition 3.1, we have:

$$\left(C \otimes_{L,\acute{e}t}^{tr} D\right) \rightarrow D \otimes_{L,\acute{e}t}^{tr} C, \tag{66}$$

Then, it is sufficient to demonstrate that $\otimes_{L,\acute{e}t}^{tr}$, preserve quasi-isomorphisms. The details can be found in [5].

Then the tensor product $\otimes_{\acute{e}t}^{tr}$, as pre-sheaf to Étale sheaves can have a homology space of zero dimension that vanishes in certain component right exact functor $\Phi(F) = R_{tr}(\mathrm{Y}) \otimes_{\acute{e}t}^{tr} F$, from the category $\mathrm{PST}(k, R)$, of pre-sheaves of $R$-modules with transfers to the category of the Étale sheaves of $R$-modules and transfers. Then every derived functor $L_n\Phi$, vanishes on $H_0(\tilde{C})$, to certain complex of Étale.

Then, all right exact functors $R_{tr}(\mathrm{Y}) \otimes_{\acute{e}t}^{tr} F$, are acyclic. This is the machinery to demonstrate the functor exactness and resolution in modules through of induce from $\otimes_{L,\acute{e}t}^{tr}$, a tensor-triangulated structure to a derived category more general that $\mathrm{D}^{-}R(\mathcal{A})$.

Also we have:

**Lemma 3.3.** Fix $Y$, and set $\Phi = R_{tr}(Y) \otimes_{\acute{e}t}^{tr}$. If $F$, is a pre-sheaf of $R$-modules with transfers such that $F_{\acute{e}t} = 0$, then $L_n\Phi(F) = 0, \ \forall n$.

## 4. Some considerations to mathematical physics

Remember that in the derived geometry we work with structures that must support $R$-modules with characterizations that should be most general to the case of singularities, where it is necessary to use irregular connections, if it is the case, for example in field theory in mathematical physics when studying the quantum field equations on a complex Riemann manifold with singularities.

---

[10] **Definition.** A pre-sheaf with transfers is a contravariant additive functor from the category $Cor_k$, to the category of abelian groups Ab.

Through the characterization of connections for derived tensor products, we search precisely generalize the connections through pre-sheaves with certain special properties, as can be the Étale sheaves.

Remember we want to generalize the field theory on spaces that admit decomposing into components that can be manageable in the complex manifolds whose complex varieties can be part of those components called motives, creating a decomposition in the derived category of its spectrum considering the functor Spec, and where solutions of the field equations are defined in a hypercohomology.[11] Likewise, this goes focused to obtain a good integrals theory (solutions) in the hypercohomology context considering the knowledge of spectral theory of the cycle sequences in motivic theory that searches the solution of the field equations even with singularities of the complex Riemann manifold.

We can demonstrate that $\otimes^{tr}_{L,ét}$, induces a tensor-triangulated structure to a derived category more general than $D^{-}R(\mathcal{A})$, as for example, $DM^{eff,-}_{ét}(k, \mathbb{Z}/m)$, which is our objective. In this case, we want geometrical motives, where this last category $DM^{eff,-}_{ét}(k, \mathbb{Z}/m)$, can be identified for the derived category $DM^{-}_{gm}(k, R)$.

We consider and fix $Y$, and the right exact functor $\Phi(F) = R_{tr}(Y) \otimes^{tr}_{ét} F$, from the category $PST(k, R)$, of pre-sheaves of $R$-modules with transfers to the category of the Étale sheaves of $R$-modules and transfers. Likewise, their left functors $L_P\Phi(F)$, are the homology sheaves of the total left derived functor $\Phi(F) = R_{tr}(Y) \otimes^{tr}_{L,ét} F$. Considering a chain complex $C$, the hypercohomology spectral sequence is:

$$E^2_{p,q} = L_p\Phi(H_q C), \tag{67}$$

then

$$L_{p+q}\Phi(C) = 0, \tag{68}$$

Then the corresponding infinite sequence is exact.

We consider $A$, and $B \in \mathcal{A}$, where $\mathcal{A}$, is a category as has been defined before.

We have the following proposition.

**Proposition 4.1.** There is equivalence between categories $\mathrm{Ab}\left(\mathrm{CRing}_{A//B}\right) \cong \mathrm{Mod}_B$.

Then a hypercohomology as given to $d\mathrm{da} = 0$, can be obtained through double functor work $A \to B \to B$, through an inclusion of a category $\mathrm{Mod}_B$, in $\mathrm{CRing}_{A//B}$. Then is had the result.

**Theorem 4.1.** The left adjoint to the inclusion functor $\mathrm{Mod}_B$, $\mathrm{CRing}_{A//B}$. is defined by $X \mapsto \Omega_{X/A} \otimes_X B$. In particular, the image of $A \to B \to B$, under this functor is $B \mapsto \Omega_{X/A}$.

The derived tensor product is a regular tensor product.

**Theorem 4.2.** The character for an adjoint lifts for a homotopically meaningful adjunction complies:

$$\mathrm{Ch}(B)_{\geq B} \leftrightarrow s\mathrm{CRing}_{A//B}, \tag{69}$$

---

[11] **Definition.** A hyperhomology or hypercohomology of a complex of objects of an abelian category is an extension of the usual homology of an object to complexes. The mechanism to give a hypercohomology is suppose that $\mathcal{A}$, is an abelian category with enough injectives and $\Phi$, a left exact functor to another abelian category $\mathcal{B}$. If $C$, is a complex of objects of $\mathcal{A}$, bounded on the left, the hypercohomology $H^i(C)$, of $C$, (for an integer $i$) is calculated as follows: take a quasi-isomorphism $\psi$ : $C \to I$, where $I$, is a complex of injective elements of $\mathcal{A}$. The hypercohomology $H^i(C)$, of $C$, is then the cohomology $H^i(\Phi(I))$, of the complex $\Phi(I)$.

Meaning that, it is an adjunction of categories, which induces an adjunction to level of homotopy categories.

We define the cotangent complex required in derived geometry and QFT.

**Definition 4.1.** The cotangent complex $L_{A/B}$, is the image of functor $A \to B \to B$, under the left functor of the Kahler differentials module $M \otimes^{L}_{R(\mathcal{A})}$, . Likewise, if $P_{\bullet} \to B$, be a free resolution then

$$L_{A/B} = \Omega_{P_{\cdot}/A} \otimes_{P_{\cdot}} B, \tag{70}$$

The cotangent complex as defined in (69) lives in the derived category $\mathrm{Mod}_B$. We observe that choosing the particular resolution of $B$, then $\Omega_{P_{\cdot}/A}$, is a co-fibrant object in the derived category $\mathrm{Mod}_{P_{\cdot}}$, which no exist distinction between the derived tensor product and the usual tensor product. Then to any representation automorphic of $G(A)$, the $G(F)/G(A)$, can be decomposed as the tensor product $\otimes_{i=1}^{n} \pi_I$. This last fall in the geometrical Langlands ramifications.

**Example 4.1.** (66) in the context of solution of field equations as $d\mathrm{da} = 0$, has solution in the hypercohomology of a spectral sequence of $\mathrm{D}^{-}R(\mathcal{A})$, (established on the infinite sequence $\ldots \to F^{n} \to 0 \to \ldots$. [6]) when its functors whose image $\Omega_{B/A}$, have as its cotangent complex the image under of the functor $L_{A/B}$, which is the functor image $A \to B \to B$, under the left derived functor of Kahler differentials.

To demonstrate this, it is necessary to define an equivalence between derived categories in the level of derived categories $D({}^{L}\mathrm{Bun}, \mathcal{D})$, and $D({}^{L}Loc, \mathcal{O})$, where geometrical motives can be risked with the corresponding moduli stack to holomorphic bundles. The integrals are those whose functors image will be in $\mathrm{Spec}_H \mathrm{SymT}(\mathrm{OP}_{{}^{L}G}(D))$, which is the variety of opers on the formal disk $D$, or neighborhood of all points in a surface $\Sigma$, in a complex Riemannian manifold [6].

## 5. Applications

As was shown, the geometrical motives required in our research are a result of embedding the derived category $\mathrm{DM}^{-}_{\mathrm{gm}}(k, R)$, (geometrical motives category) in the $\mathrm{DM}^{eff,-}_{\acute{e}t}(k, \mathbb{Z}/m)$, considering the category of smooth schemes on the field $k$.

We consider the following functors. For each $F \in \mathrm{D}^{-}\left(\mathrm{Sh}^{\mathrm{Nis}}(Cor(k))\right)$, there is $L^{A^1} F \in \mathrm{D}^{\mathrm{eff}}_{-}(k)$, the resulting functor is:

$$L^{A^1} : \mathrm{D}^{-}\left(\mathrm{Sh}^{\mathrm{Nis}}(Cor(k))\right) \to \mathrm{D}^{\mathrm{eff}}_{-}(k), \tag{71}$$

which is exact and left-adjoint to the inclusion

$$\mathrm{D}^{\mathrm{eff}}_{-}(k) \to \mathrm{D}^{-}\left(\mathrm{Sh}^{\mathrm{Nis}}(Cor(k))\right), \tag{72}$$

Also the functor (70) descends to an equivalence of triangulated categories. This is very useful to make $\mathrm{D}^{\mathrm{eff}}_{-}(k)$, into a tensor category as follows. We consider the Nisnevich sheaf $\mathbb{Z}_{tr}(k)$, with transfer $tr : \mathrm{Y} \to c(\mathrm{Y}, \mathrm{X})$. We define

$$\mathbb{Z}_{tr}(k) \otimes \mathbb{Z}_{tr}(k) := \mathbb{Z}_{tr}(\mathrm{X} \times_k \mathrm{Y}), \tag{73}$$

Then it can be demonstrated that the operation realized in (70) can be extended to give $\mathrm{D}^{-}\left(\mathrm{Sh}^{\mathrm{Nis}}(Cor(k))\right)$, with the structure of a triangulated tensor category.

Then the functor $L_{A^1}$, induces a tensor operation on $\mathrm{D}^{-}_{A^1}\left(\mathrm{Sh}^{\mathrm{Nis}}(Cor(k))\right)$, making that $\mathrm{D}^{-}_{A^1}\left(\mathrm{Sh}^{\mathrm{Nis}}(Cor(k))\right)$ also a triangulated tensor category. Likewise, explicitly in $\mathrm{DM}^{\mathrm{eff}}_{-}(k)$, this gives us the functor

$$m : Sm_k \to \mathrm{DM}^{\mathrm{eff}}_{-}(k), \tag{74}$$

defined by

$$m(\mathrm{X}) := C^{\mathrm{Sus}}(\mathrm{Z}_{tr}(\mathrm{X})), \tag{75}$$

where we have the formula

$$m(\mathrm{X}\times_k \mathrm{Y}) = m(\mathrm{X}) \otimes m(\mathrm{Y}), \tag{76}$$

If we consider the embedding theorem, then we can establish the following triangulated scheme

$$\begin{array}{c} Sm_k \to \mathrm{DM}^{\mathrm{eff}}_{\mathrm{gm}}(k) \\ m \searrow \quad \updownarrow \mathrm{Id} \\ \mathrm{DM}^{\mathrm{eff}}_{\mathrm{gm}}(k) \end{array} \tag{77}$$

which has implications in the geometrical motives applied to bundle of geometrical stacks in mathematical physics.

**Theorem 5.1 (F. Bulnes).** Suppose that $M$, is a complex Riemannian manifold with singularities. Let $X$, and $Y$, be smooth projective varieties in $M$ [12]. We know that solutions of the field equations $d\mathrm{da} = 0$, are given in a category $\mathrm{Spec}(Sm_{\mathrm{k}})$, (see Example 4). Context Solutions of the quantum field equations for $d\mathrm{da} = 0$, are defined in hyper-cohomology on $\mathbb{Q}$- coefficients from the category $Sm_k$, defined on a numerical field $k$, considering the derived tensor product $\otimes^{tr}_{ét}$, of pre-sheaves. Then the following triangulated tensor category scheme is true and commutative:

$$\begin{array}{ccc} & \mathrm{DQFT} & \\ i\swarrow & & \searrow F \\ \mathrm{MD}_{\mathrm{gm}}(\mathbb{Q}) & \leftrightarrow & \mathrm{MD}(\mathcal{O}_{\mathrm{Y}}) \end{array} \tag{78}$$

The category $\mathrm{DM}^{\mathrm{eff}}_{\mathrm{gm}}(k, R)$, has a tensor structure and the tensor product of its motives is as defined in (75) $m(\mathrm{X}) \otimes m(\mathrm{Y}) = m(\mathrm{X} \times \mathrm{Y})$.

Triangulated category of geometrical motives $\mathrm{DM}_{\mathrm{gm}}(k, R)$, or written simply as $\mathrm{DM}_{\mathrm{gm}}(k)$, is defined formally inverting the functor of the Tate objects[13] (are objects of a motivic category called Tannakian category) $\mathbb{Z}\,(1)$, to be image of the complex

[12] Singular projective varieties useful in quantization process of the complex Riemannian manifold. The quantization condition compact quantizable Käehler manifolds can be embedded into projective space.

[13] Let $\mathrm{MT}(\mathbb{Z})$, denote the category of mixed Tate motives unramified over $\mathbb{Z}$. It is a Tannakian category with Galois group $\mathrm{Gal}_{\mathrm{MT}}$.
The inverting of the objects $- \otimes \mathbb{Z}.(1)$.
Remember that a scheme is a mathematical structure that enlarges the concept of algebraic variety in several forms, such as taking account of multiplicities. The schemes can to be of a same algebraic variety different and allowing "varieties" defined over any commutative ring. In many cases, the family of all varieties of a type can be viewed as a variety or scheme, known as a moduli space.

$[\mathbb{P}^1] \to [\mathrm{Spec}(k)]$, where the motive in degree $p = 2, 3$, will be $m(p) = m \otimes \mathbb{Z}(1)^{\otimes p}$, or to any motive $m \in \mathrm{DM}^{\mathrm{eff}}_{\mathrm{gm}}(k), \forall p \in \mathbb{N}$.

Likewise, the important fact is that the canonical functor $\mathrm{DM}^{\mathrm{eff}}_{\mathrm{gm}}(k), \to \mathrm{DM}_{\mathrm{gm}}(k)$, is full embedding [7]. Therefore we work in the category $\mathrm{DM}_{\mathrm{gm}}(k)$.

Likewise, for $X$, and $Y$, smooth projective varieties and for any integer $i$, there exists an isomorphism:

$$\mathrm{Hom}_{\mathrm{DM}^{\mathrm{eff}}_{\mathrm{gm}}(k)}(m(\mathrm{X}), m(\mathrm{Y})(i)[2i]) \cong A^{m+i}(\mathrm{X} \times \mathrm{Y}), \qquad m = \dim \mathrm{Y}, \tag{79}$$

We demonstrate the Theorem 5.1.

*Proof*. $\forall\, \mathrm{X} \in Sm_k$, the category $Sm_k$, extends to a pseudo-tensor equivalence of cohomological categories over motives on $k$, that is to say, $\mathrm{MM}(k)$, is the image of functors

$$\mathrm{DM}^{\mathrm{eff}}(k) \to \mathrm{DM}_{\mathrm{gm}}(k), \tag{80}$$

which is an equivalence of the underlying triangulated tensor categories.

On the other hand, the category DQFT can be defined for the motives in a hypercohomology from the category $Sm_k$, defined as:

$$\mathrm{Hom}_{\mathrm{DM}^{\mathrm{eff}}_{\mathrm{gm}}(k)}(m(\mathrm{X}), \mathbb{Q}(q)[\mathrm{p}]) \cong H^{\bullet}_{Nis}(\mathrm{X}, \mathbb{Q}, (q)) = \mathbb{H}^{p,q}(\mathrm{X}_{Nis}, \mathbb{Q}, (q)), \tag{81}$$

which comes from the hypercohomology

$$H^{p,q}_L(\mathrm{X}, \mathbb{Q}) = \mathbb{H}^{p,q}(\mathrm{X}, \mathbb{Q}), \tag{82}$$

We observe that if a Zariski sheaf of $\mathbb{Q}$-modules with transfers $F$, is such that $F = H^q C$, for all $C$, a complex defined on $\mathbb{Q}(q)$-modules (being a special case when $C = \mathbb{Q}(q)$), where the cohomology groups of the isomorphism $H^p_{\acute{e}t}(\mathrm{X}, F_{\acute{e}t}) \cong H^p_{Nis}(\mathrm{X}, F_{Nis})$, can be vanished for $p > \dim(\mathrm{Y})$.

Then survives a hypercohomology $\mathbb{H}^q(\mathrm{X}, \mathbb{Q})$. If we consider $\mathrm{Spec}(Sm_{\mathrm{k}})$, we can to have the quantum version of this hyper-cohomology with an additional work on moduli stacks of the category $\mathrm{Mod}_B$, in a study on equivalence between derived categories in the level of derived categories $D({}^L\mathrm{Bun}, \mathcal{D})$, and $D({}^L Loc, \mathcal{O})$, where geometrical motives can be risked with the corresponding moduli stack to holomorphic bundles[14].

For other way, with other detailed work of quasi-coherent sheaves [6] we can to obtain the category $\mathrm{MO}_{\mathcal{O}}(\mathrm{Y})$. The functors are constructed using the Mukai-Fourier transforms. ■

[14] We consider the functor $F$, defined as:
where $\mathcal{K}(F^r)$, the kernel space of the functor $F^r$, is the functor that induces the equivalence $\mathrm{Mod}_T(D(\mathrm{X}\times_{\mathrm{Y}}\mathrm{X})) \cong {}^{\perp}\mathcal{K}(F^r)$, and the operator $T = F^r \circ F$, acting on category $D(\mathrm{X}\times_{\mathrm{Y}}\mathrm{X})$.

## Acknowledgements

Posdoctoral research was supported by State of Mexico Council of Scientific Research COMECYT-077/111/21.

## Author details

Francisco Bulnes
Research Department in Mathematics and Engineering, TESCHA, IINAMEI (Investigación Internacional en Matemáticas Avanzadas e Ingeniería), Chalco, Mexico

*Address all correspondence to: francisco.bulnes@tesch.edu.mx

## References

[1] Adams JF. Stable homotopy and generalised homology. In: Chicago Lectures in Mathematics, MR 53 #6534. Chicago, IL: University of Chicago Press; 1974

[2] Mazza C, Voevodsky V, Weibel C, editors. Lecture Notes on Motivic Cohomology. Vol. 2. Cambridge, MA, USA: AMS Clay Mathematics Institute; 2006

[3] Aubry M. Homotopy Theory and Models. Boston, MA: Birkhäuser; 1995

[4] Milne JS. Étale Cohomology. Princeton, NJ: Princeton University Press; 1980. ISBN: 0-691-08238-3

[5] Bosch S. Algebraic Geometry and Commutative Algebra. NY, USA: Springer; 2013

[6] Bulnes F. Extended d-cohomology and integral transforms in derived geometry to QFT-equations solutions using Langlands correspondences. Theoretical Mathematics and Applications. 2017;7(2):51-62

[7] Voevodsky V. Triangulated categories of motives over a field. In: Cycles, Transfers, and Motivic Homology Theories, Vol. 143 of Annals of Mathematics Studies. USA: Princeton University Press; 2000. pp. 188-238

Chapter 7

# Higher-Order Kinematics in Dual Lie Algebra

*Daniel Condurache*


## Abstract

In this chapter, using the ring properties of dual number algebra, vector and tensor calculus, a computing method for the higher-order acceleration vector field properties in general rigid body motion is proposed. The higher-order acceleration field of a rigid body in a general motion is uniquely determined by higher-order time derivative of a dual twist. For the relative kinematics of rigid body motion, equations that allow the determination of the higher-order acceleration vector field are given, using an exponential Brockett-like formula in the dual Lie algebra. In particular cases, the properties for velocity, acceleration, jerk, and jounce fields are given. This approach uses the isomorphism between the Lie algebra of the rigid displacements $se(3)$, of the Special Euclidean group, $\mathbb{SE}_3$, and the Lie algebra of dual vectors. The results are coordinate free and in a closed form.


**Keywords:** higher-order kinematics, dual algebra, lie group

## 1. Introduction

The kinematic analysis of multibody systems has been traditionally considered as the determination of the positions, velocities, accelerations, jerks and jounces of their constitutive members. This is an old field with a long history, which has attracted the attention of mathematicians and engineers. Michel Chasles discovered (1834) that any rigid body displacement is equivalent to a screw displacement [1]. Screw theory is an efficient mathematical tool for the study of spatial kinematics. The pioneering work of Ball [2], the treatises of Hunt [3], and Phillips [4] and the multitude of contributions appearing in the literature are evidence of this. The isomorphism between screw theory and the Lie algebra, se(3), of the Special Euclidean group, $\mathbb{SE}_3$, provide with a wealth of results and techniques from modern differential geometry and Lie group theory [5–9].

A kinematic mapping relates the motion of a rigid body to the joint motions of a kinematic chain. Its time derivatives yield the twist, acceleration, jerk and jounce etc. of the body. Time derivatives of the twists of members in a kinematic chain and derivatives of screws are essential operations in kinematics. Recognizing the Lie group nature of rigid body motions, and correspondingly the Lie algebra nature of screws, Karger [5], Rico et al. [6], Lerbet [7] and Müller [8, 9] derived closed form expressions of higher-order time derivatives of twist.

In this chapter, using the tensor calculus and the dual numbers algebra, a new computing method for studying the higher-order accelerations field properties is proposed in the case of the general rigid body motion. For the spatial kinematic chains, equations that allow the determination of the $n^{th}$ order accelerations field are given, using a Brockett-like formula. The crucial observation is that the $n^{th}$ order time derivative of twist of the terminal body in a kinematic chain can be determined by propagating the $k^{th}$ order time derivative of twists of the bodies in the chain, for $k = \overline{0,n}$. The results are coordinate-free and in a closed form.

## 2. Theoretical consideration on rigid body motion

The general framework of this chapter is a rigid body that moves with respect to a fixed reference frame $\{\mathfrak{R}^0\}$. Consider another reference frame $\{\mathfrak{R}\}$ originated in a point $Q$ that moves together with the rigid body. Let $\boldsymbol{\rho}_Q$ denote the position vector of point $Q$ with respect to frame $\{\mathfrak{R}^0\}$, $\mathbf{v}_Q$ its absolute velocity and $\mathbf{a}_Q$ its absolute acceleration.

Then the vector parametric equation of motion is:

$$\boldsymbol{\rho} = \boldsymbol{\rho}_Q + \mathbf{R}\mathbf{r} \tag{1}$$

where $\boldsymbol{\rho}$ represents the absolute position of a generic point $P$ of the rigid body with respect to $\{\mathfrak{R}^0\}$ and $\mathbf{R} = \mathbf{R}(t)$ is an orthogonal proper tensorial function in $\mathbf{SO}_3^{\mathbb{R}}$. Vector $\mathbf{r}$ is constant and it represents the relative position vector of the arbitrary point $P$ with respect to $\{\mathfrak{R}\}$.

The results of this section succinctly present the velocity and acceleration vector field in rigid body motion. These results lead to the generalization presented in the next section.

With the denotations that were introduced, the vector fields of velocities and accelerations are described by:

$$\begin{cases} \mathbf{v} - \mathbf{v}_Q = \dot{\mathbf{R}}\mathbf{R}^T(\boldsymbol{\rho} - \boldsymbol{\rho}_Q) \\ \mathbf{a} - \mathbf{a}_Q = \ddot{\mathbf{R}}\mathbf{R}^T(\boldsymbol{\rho} - \boldsymbol{\rho}_Q) \end{cases} \tag{2}$$

Tensors:

$$\begin{cases} \boldsymbol{\Phi}_1 = \dot{\mathbf{R}}\mathbf{R}^T \\ \boldsymbol{\Phi}_2 = \ddot{\mathbf{R}}\mathbf{R}^T \end{cases} \tag{3}$$

represent the **velocity tensor** respectively the **acceleration tensor**. Tensor $\boldsymbol{\Phi}_1 = \tilde{\boldsymbol{\omega}} \in \mathbf{so}_3^{\mathbb{R}}$ is the skew-symmetric tensor associated to the instantaneous angular velocity $\boldsymbol{\omega} \in \mathbf{V}_3^{\mathbb{R}}$. Tensor $\boldsymbol{\Phi}_2 = \tilde{\boldsymbol{\omega}}^2 + \tilde{\boldsymbol{\varepsilon}}$, where $\boldsymbol{\varepsilon} = \dot{\boldsymbol{\omega}}$ is the instantaneous angular acceleration of the rigid body. One may remark that vectors:

$$\begin{cases} \mathbf{a}_1 = \mathbf{v} - \boldsymbol{\Phi}_1\boldsymbol{\rho} = \mathbf{v}_Q - \boldsymbol{\Phi}_1\boldsymbol{\rho}_Q \\ \mathbf{a}_2 = \mathbf{a} - \boldsymbol{\Phi}_2\boldsymbol{\rho} = \mathbf{a}_Q - \boldsymbol{\Phi}_2\boldsymbol{\rho}_Q \end{cases} \tag{4}$$

do not depend on the choice of point $P$ of the rigid body. They are called the **velocity invariant** respectively the **acceleration invariant** (at a given moment of time).

### 2.1 The velocity field in rigid body motion

It is described by:

$$\mathbf{v} - \mathbf{v}_Q = \mathbf{\Phi}_1(\boldsymbol{\rho} - \boldsymbol{\rho}_Q) \tag{5}$$

The instantaneous angular velocity $\boldsymbol{\omega}$ of the rigid body may be determined as $\boldsymbol{\omega} = \text{vect}\mathbf{\Phi}_1$. The major property that may be highlighted from Eq. (4) is that the velocity of a given point of the rigid may be computed when knowing the velocity tensor $\mathbf{\Phi}_1$ and the velocity invariant $\mathbf{a}_1$:

$$\mathbf{v} = \mathbf{a}_1 + \mathbf{\Phi}_1\boldsymbol{\rho} \tag{6}$$

### 2.2 The acceleration field in rigid body motion

It is described by

$$\mathbf{a} - \mathbf{a}_Q = \mathbf{\Phi}_2(\boldsymbol{\rho} - \boldsymbol{\rho}_Q) \tag{7}$$

The absolute acceleration of a given point of the rigid body may be computed when knowing the acceleration tensor $\mathbf{\Phi}_2$ and the acceleration invariant $\mathbf{a}_2$:

$$\mathbf{a} = \mathbf{a}_2 + \mathbf{\Phi}_2\boldsymbol{\rho} \tag{8}$$

The instantaneous angular acceleration of the rigid body may be determined as:

$$\boldsymbol{\varepsilon} = \text{vect}\mathbf{\Phi}_2 \tag{9}$$

The determinant of tensor $\mathbf{\Phi}_2$ is (see [10]): $\det\mathbf{\Phi}_2 = -(\boldsymbol{\omega} \times \boldsymbol{\varepsilon})^2$. It follows that if $\boldsymbol{\omega} \times \boldsymbol{\varepsilon} \neq \mathbf{0}$, then tensor $\mathbf{\Phi}_2$.
is invertible and its inverse is (see [10]):

$$\mathbf{\Phi}_2^{-1} = \frac{1}{(\boldsymbol{\omega} \times \boldsymbol{\varepsilon})^2}\left[\boldsymbol{\varepsilon} \otimes \boldsymbol{\varepsilon} + (\boldsymbol{\omega} \otimes \boldsymbol{\omega})^2 - \widetilde{\tilde{\boldsymbol{\omega}}^2\boldsymbol{\varepsilon}}\right] \tag{10}$$

It follows that if tensor $\mathbf{\Phi}_2$ is non-singular, then for an arbitrary given acceleration $\mathbf{a}$ we may find a point of the rigid that has this acceleration. Its absolute position is given by (see also Eq. (8)):

$$\boldsymbol{\rho} = \mathbf{\Phi}_2^{-1}(\mathbf{a} - \mathbf{a}_2) \tag{11}$$

Particularly, if $\mathbf{\Phi}_2$ is non-singular, then there exists a point $G$ of zero acceleration, named the **acceleration center**. Its absolute position vector is given by:

$$\boldsymbol{\rho}_G = -\mathbf{\Phi}_2^{-1}\mathbf{a}_2 \tag{12}$$

## 3. The vector field of the $n^{th}$ order accelerations

This section extends some of the previous considerations to the case of the $n^{th}$ order accelerations. We define the $n^{th}$ order acceleration of a point as:

$$\mathbf{a}_{\boldsymbol{\rho}}^{[n]} \stackrel{\text{def}}{=} \frac{d^n}{dt^n}\boldsymbol{\rho}, n \geq 1 \tag{13}$$

For $n = 1$, it represents the velocity, and for $n = 2$, the acceleration. By derivation with respect to time successively in Eq. (2), it follows that:

$$\mathbf{a}_{\boldsymbol{\rho}}^{[n]} - \mathbf{a}_Q^{[n]} = \mathbf{R}^{(n)}\mathbf{R}^T(\boldsymbol{\rho} - \boldsymbol{\rho}_Q), \text{where } \mathbf{R}^{(n)} \stackrel{\text{def}}{=} \frac{d^n}{dt^n}\mathbf{R} \tag{14}$$

We define:

$$\boldsymbol{\Phi}_n \stackrel{\text{def}}{=} \mathbf{R}^{(n)}\mathbf{R}^T \tag{15}$$

the ***n**<sup></sup>* $^{th}$ **order acceleration tensor** in rigid body motion. A vector invariant is immediately highlighted from Eq. (14) with the denotation (15). Vector:

$$\mathbf{a}_n = \mathbf{a}_{\boldsymbol{\rho}}^{[n]} - \boldsymbol{\Phi}_n\boldsymbol{\rho} = \mathbf{a}_Q^{[n]} - \boldsymbol{\Phi}_n\boldsymbol{\rho}_Q \tag{16}$$

does not depend on the choice of the point of the rigid body for which the acceleration $\mathbf{a}^{[n]}$ is computed. Vector $\mathbf{a}_n$ is named the **invariant vector** of the $\boldsymbol{n^{th}}$ **order accelerations**. Then Eq. (7) may be generalized as it follows:

$$\mathbf{a}_{\boldsymbol{\rho}}^{[n]} - \mathbf{a}_Q^{[n]} = \boldsymbol{\Phi}_n(\boldsymbol{\rho} - \boldsymbol{\rho}_Q) \tag{17}$$

The next Theorem gives the fundamental properties of the vector field of the $n^{th}$ order accelerations.

**Theorem 1.** *In the rigid body motion, at a moment of time t, there exist tensor* $\boldsymbol{\Phi}_n$ *defined by Eq. (15) and vector* $\mathbf{a}_n$ *such as:*

$$\begin{aligned} \mathbf{a}_{\boldsymbol{\rho}}^{[n]} - \mathbf{a}_Q^{[n]} &= \boldsymbol{\Phi}_n(\boldsymbol{\rho} - \boldsymbol{\rho}_Q) \\ \mathbf{a}_n = \mathbf{a}_{\boldsymbol{\rho}}^{[n]} - \boldsymbol{\Phi}_n\boldsymbol{\rho} &= \mathbf{a}_Q^{[n]} - \boldsymbol{\Phi}_n\boldsymbol{\rho}_Q \end{aligned} \tag{18}$$

for any point $P$ of the rigid body with the absolute position defined by vector $\boldsymbol{\rho}$.

**Remark 1.** *Given the absolute position of a point of the rigid body and knowing* $\boldsymbol{\Phi}_n$ *and* $\mathbf{a}_n$*, its acceleration is computed from:*

$$\mathbf{a}_{\boldsymbol{\rho}}^{[n]} = \mathbf{a}_n + \boldsymbol{\Phi}_n\boldsymbol{\rho} \tag{19}$$

**Remark 2.** *Tensor* $\boldsymbol{\Phi}_n$ *and vector* $\mathbf{a}_n$ *generalize the notions of velocity/acceleration tensor respectively velocity/acceleration invariant. They are fundamental in the study of the vector field of the* $n^{th}$ *order accelerations. The recursive formulas for computing* $\boldsymbol{\Phi}_n$ *and* $\mathbf{a}_n$ *are:*

$$\begin{cases} \boldsymbol{\Phi}_{n+1} = \dot{\boldsymbol{\Phi}}_n + \boldsymbol{\Phi}_n\boldsymbol{\Phi}_1 \\ \mathbf{a}_{n+1} = \dot{\mathbf{a}}_n + \boldsymbol{\Phi}_n\mathbf{a}_1 \end{cases}, n \geq 1, \text{where } \boldsymbol{\Phi}_1 = \tilde{\boldsymbol{\omega}}, \mathbf{a}_1 = \mathbf{v}_Q - \boldsymbol{\Phi}_1\boldsymbol{\rho}_Q \tag{20}$$

**Remark 3.** *One may remark that from Eq. (20) it follows by direct computation:*

$$\begin{cases} \boldsymbol{\Phi}_n = \boldsymbol{\Phi}_{n-1}\boldsymbol{\Phi}_1 + \left(\dfrac{d^{n-1}}{dt^{n-1}}\boldsymbol{\Phi}_1\right) + \displaystyle\sum_{k=1}^{n-2}\left[\frac{d^k}{dt^{n-1}}(\boldsymbol{\Phi}_{n-k-1}\boldsymbol{\Phi}_1)\right] \\ \mathbf{a}_n = \boldsymbol{\Phi}_{n-1}\mathbf{a}_1 + \left(\dfrac{d^{n-1}}{dt^{n-1}}\mathbf{a}_1\right) + \displaystyle\sum_{k=1}^{n-2}\left[\frac{d^k}{dt^{n-1}}(\boldsymbol{\Phi}_{n-k-1}\mathbf{a}_1)\right] \end{cases}, n \geq 3 \tag{21}$$

*DOI: http://dx.doi.org/10.5772/intechopen.91779*

**Remark 4.** *By defining the $n^{th}$ order instantaneous $n^{th}$ order angular acceleration of the rigid body $\boldsymbol{\varepsilon}^{[n]} \stackrel{\text{def}}{=} \frac{d^{n-1}}{dt^{n-1}}\boldsymbol{\omega}$, it follows from Eq. (21) that its associated skew-symmetric tensor may be expressed as $\tilde{\boldsymbol{\varepsilon}}^{[n]} = \frac{d^{n-1}}{dt^{n-1}}\boldsymbol{\Phi}_1$. The expression of the instantaneous $n^{th}$ order angular acceleration is:*

$$\boldsymbol{\varepsilon}^{[n]} = vect\left\{\boldsymbol{\Phi}_n - \boldsymbol{\Phi}_{n-1}\boldsymbol{\Phi}_1 - \sum_{k=1}^{n-2}\left[\frac{d^k}{dt^k}(\boldsymbol{\Phi}_{n-k-1}\boldsymbol{\Phi}_1)\right]\right\}, n \geq 3 \tag{22}$$

### 3.1 Homogenous matrix approach to the field of $n^{th}$ order accelerations

The set of affine maps, $g : \mathbf{V}_3 \to \mathbf{V}_3, g(\mathbf{u}) = \mathbf{R}\mathbf{u} + \mathbf{w}$, where $\mathbf{R}$ is an orthogonal proper tensor and $\mathbf{w}$ a vector in $\mathbf{V}_3$ is a group under composition and it is called *the group of direct affine isometries* or *rigid motions* and it is denoted $S\mathbb{E}_3$. Any rigid finite motion may be described by such a map. Tensor $\mathbf{R}$ models the rotation of the considered rigid body and vector $\mathbf{w}$ its translation. An affine map from $S\mathbb{E}_3$ may be represented with a $4 \times 4$ square matrix:

$$g = \begin{bmatrix} \mathbf{R} & \mathbf{w} \\ \mathbf{0} & 1 \end{bmatrix} \tag{23}$$

One may remark that the following relations hold true:

$$\begin{cases} \begin{bmatrix} \mathbf{R}_1 & \mathbf{w}_1 \\ \mathbf{0} & 1 \end{bmatrix}\begin{bmatrix} \mathbf{R}_2 & \mathbf{w}_2 \\ \mathbf{0} & 1 \end{bmatrix} = \begin{bmatrix} \mathbf{R}_1\mathbf{R}_2 & \mathbf{R}_1\mathbf{w}_2 + \mathbf{w}_1 \\ \mathbf{0} & 1 \end{bmatrix} \\ \begin{bmatrix} \mathbf{R} & \mathbf{w} \\ \mathbf{0} & 1 \end{bmatrix}^{-1} = \begin{bmatrix} \mathbf{R}^T & -\mathbf{R}^T\mathbf{w} \\ \mathbf{0} & 1 \end{bmatrix} \end{cases} \tag{24}$$

We may extend now $S\mathbb{E}_3$ to $S\mathbb{E}_3^{\mathbb{R}}$, the set of the functions with the domain $\mathbb{R}$ and the range $S\mathbb{E}_3$. The parametric vector equation of the rigid body motion (1) may be rewritten with the help of a homogenous matrix function in $S\mathbb{E}_3^{\mathbb{R}}$ like it follows:

$$\begin{bmatrix} \boldsymbol{\rho} \\ 1 \end{bmatrix} = \begin{bmatrix} \mathbf{R} & \boldsymbol{\rho}_Q \\ \mathbf{0} & 1 \end{bmatrix}\begin{bmatrix} \mathbf{r} \\ 1 \end{bmatrix} \tag{25}$$

From Eq. (25), it follows that:

$$\begin{bmatrix} \dot{\boldsymbol{\rho}} \\ 0 \end{bmatrix} = \begin{bmatrix} \dot{\mathbf{R}} & \dot{\boldsymbol{\rho}}_Q \\ \mathbf{0} & 0 \end{bmatrix}\begin{bmatrix} \mathbf{r} \\ 1 \end{bmatrix} = \begin{bmatrix} \dot{\mathbf{R}} & \dot{\boldsymbol{\rho}}_Q \\ \mathbf{0} & 0 \end{bmatrix}\begin{bmatrix} \mathbf{R}^T & -\mathbf{R}^T\boldsymbol{\rho}_Q \\ \mathbf{0} & 1 \end{bmatrix}\begin{bmatrix} \boldsymbol{\rho} \\ 1 \end{bmatrix} \tag{26}$$

and by making the computations and taking into account Eqs. (3) and (4) it follows that:

$$\begin{bmatrix} \dot{\boldsymbol{\rho}} \\ 0 \end{bmatrix} = \begin{bmatrix} \boldsymbol{\Phi}_1 & \mathbf{a}_1 \\ \mathbf{0} & 0 \end{bmatrix}\begin{bmatrix} \boldsymbol{\rho} \\ 1 \end{bmatrix} \tag{27}$$

By using the previous considerations, it follows that Eq. (25) may be extended like:

$$\begin{bmatrix} \mathbf{a}_{\rho}^{[n]} \\ 0 \end{bmatrix} = \begin{bmatrix} \mathbf{\Phi}_n & \mathbf{a}_n \\ \mathbf{0} & 0 \end{bmatrix} \begin{bmatrix} \boldsymbol{\rho} \\ 1 \end{bmatrix} \tag{28}$$

Eq. (28) represents a unified form of describing the vector field of the $n^{th}$ order accelerations in rigid body motion. The matrix:

$$\mathbf{\Psi}_n = \begin{bmatrix} \mathbf{\Phi}_n & \mathbf{a}_n \\ \mathbf{0} & 0 \end{bmatrix} \tag{29}$$

contains both the $n^{th}$ order acceleration tensor $\mathbf{\Phi}_n$ and the vector invariant $\mathbf{a}_n$. Eqs. (20) may be put in a compact form:

$$\mathbf{\Psi}_{n+1} = \dot{\mathbf{\Psi}}_n + \mathbf{\Psi}_n\mathbf{\Psi}_1, n \geq 1 \tag{30}$$

If follows that $\mathbf{\Psi}_n$ may be written as:

$$\mathbf{\Psi}_n = \mathbf{\Psi}_{n-1}\mathbf{\Psi}_1 + \left(\frac{d^{n-1}}{dt^{n-1}}\mathbf{\Psi}_1\right) + \sum_{k=1}^{n-2}\left[\frac{d^k}{dt^k}(\mathbf{\Psi}_n\mathbf{\Psi}_1)\right], n \geq 3 \tag{31}$$

## 4. Symbolic calculus of higher-order kinematics invariants

We will present a method for the symbolic calculation of higher-order kinematics invariants for rigid motion.

Let be $\mathbf{a}_n$ and $\mathbf{\Phi}_n$, $n \in \mathbb{N}$ vector invariant, respectively, tensor invariant for the $n^{th}$ order accelerations fields. We denote by

$$\mathbf{\Psi}_n = \begin{bmatrix} \mathbf{\Phi}_n & \mathbf{a}_n \\ \mathbf{0} & 0 \end{bmatrix} \tag{32}$$

and we have the following relationship of recurrence:

$$\begin{aligned} \mathbf{\Psi}_{n+1} &= \dot{\mathbf{\Psi}}_n + \mathbf{\Psi}_n\mathbf{\Psi}_1, n \in \mathbb{N} \\ \mathbf{\Psi}_1 &= \begin{bmatrix} \tilde{\boldsymbol{\omega}} & \mathbf{v} \\ \mathbf{0} & 0 \end{bmatrix} \end{aligned} \tag{33}$$

The pair of vectors $(\boldsymbol{\omega}, \mathbf{v})$ is also known as *the spatial twist of rigid body*.

Let be $\boldsymbol{\mathcal{A}}$ the matrix ring

$$\boldsymbol{\mathcal{A}} = \left\{ \mathbf{A} \in \mathcal{M}_{3\times3}(\mathbb{R}) \middle| \mathbf{A} = \begin{bmatrix} \mathbf{\Phi} & \mathbf{a} \\ \mathbf{0} & 0 \end{bmatrix}; \mathbf{\Phi} \in \mathrm{L}(\mathbf{V}_3, \mathbf{V}_3), \mathbf{a} \in \mathbf{V}_3 \right\} \tag{34}$$

and $\boldsymbol{\mathcal{A}}[\mathrm{X}]$ the set of polynomials with coefficients in the non-commutative ring $\boldsymbol{\mathcal{A}}$. A generic element of $\boldsymbol{\mathcal{A}}[\mathrm{X}]$ has the form

$$\mathrm{P(X)} = \mathbf{A}_0\mathrm{X}^m + \mathbf{A}_1\mathrm{X}^{m-1} + \ldots + \mathbf{A}_{m-1}\mathrm{X} + \mathbf{A}_m, \mathbf{A}_k \in \boldsymbol{\mathcal{A}}, k = \overline{0, m} \tag{35}$$

**Theorem 2.** *There is a unique polynomial* $\mathbf{P}_n \in \boldsymbol{\mathcal{A}}[\mathrm{X}]$ *such that* $\mathbf{\Psi}_\mathrm{n}$ *will be written as*

$$\mathbf{\Psi}_n = \mathbf{P}_n(\mathbf{D})\mathbf{\Psi}_1, n \in \mathbb{N} \tag{36}$$

*where* $\mathbf{D} = \frac{d}{dt}$ *is the operator of time derivative.*

*Proof*: Taking into account Eqs. (36) and (33) we will have the following relationship of recurrence for $\mathbf{P}_n(\mathbf{D})$:

$$\begin{cases} \mathbf{P}_{n+1} = \mathbf{D}\mathbf{P}_n + \mathbf{P}_n(\mathbf{\Psi}_1) \\ \mathbf{P}_0 = \mathbf{I} \end{cases} \tag{37}$$

Since $\mathbf{\Psi}_1 = \begin{bmatrix} \tilde{\boldsymbol{\omega}} & \mathbf{v} \\ \mathbf{0} & 0 \end{bmatrix}$ it follows the next outcome.

**Theorem 3.** *There is a unique polynomial with the coefficients in the non-commutative ring* $\mathbf{L}(\mathbf{V}_3, \mathbf{V}_3)$ *such that the vector respectively the tensor invariants of the* $n^{th}$ *order accelerations will be written as*

$$\begin{matrix} \mathbf{a}_n = \mathbf{P}_n\mathbf{v} \\ \mathbf{\Phi}_n = \mathbf{P}_n\tilde{\boldsymbol{\omega}} \end{matrix}, n \in \mathbb{N}^* \tag{38}$$

*where* $\mathbf{P}_n$ *fulfills the relationship of recurrence*

$$\begin{cases} \mathbf{P}_{n+1} = \mathbf{D}\mathbf{P}_n + \mathbf{P}_n(\tilde{\boldsymbol{\omega}}), n \in \mathbb{N}^* \\ \mathbf{P}_1 = \mathbf{I} \end{cases} \tag{39}$$

It follows

$$\begin{gathered} \mathbf{P}_2 = \mathbf{D} + \tilde{\boldsymbol{\omega}} \\ \mathbf{P}_3 = \mathbf{D}^2 + \tilde{\boldsymbol{\omega}}\mathbf{D} + 2\dot{\tilde{\boldsymbol{\omega}}} + \tilde{\boldsymbol{\omega}}^2 \\ \mathbf{P}_4 = \mathbf{D}^3 + \tilde{\boldsymbol{\omega}}\mathbf{D}^2 + \left(3\dot{\tilde{\boldsymbol{\omega}}} + \tilde{\boldsymbol{\omega}}^2\right)\mathbf{D} + 3\ddot{\tilde{\boldsymbol{\omega}}} + 3\dot{\tilde{\boldsymbol{\omega}}}\tilde{\boldsymbol{\omega}} + 2\tilde{\boldsymbol{\omega}}\dot{\tilde{\boldsymbol{\omega}}} + \tilde{\boldsymbol{\omega}}^3 \end{gathered} \tag{40}$$

Thus, it follows:

- the velocity field invariants

$$\begin{cases} \mathbf{a}_1 = \mathbf{v} \\ \mathbf{\Phi}_1 = \tilde{\boldsymbol{\omega}} \end{cases} \tag{41}$$

- the acceleration field invariants

$$\begin{cases} \mathbf{a}_2 = \dot{\mathbf{v}} + \tilde{\boldsymbol{\omega}}\mathbf{v} \\ \mathbf{\Phi}_2 = \dot{\tilde{\boldsymbol{\omega}}} + \tilde{\boldsymbol{\omega}}^2 \end{cases} \tag{42}$$

- jerk field invariants

$$\begin{cases} \mathbf{a}_3 = \ddot{\mathbf{v}} + \tilde{\boldsymbol{\omega}}\mathbf{v} + 2\dot{\tilde{\boldsymbol{\omega}}}\mathbf{v} + \tilde{\boldsymbol{\omega}}^2\mathbf{v} \\ \mathbf{\Phi}_3 = \ddot{\tilde{\boldsymbol{\omega}}} + \tilde{\boldsymbol{\omega}}\dot{\tilde{\boldsymbol{\omega}}} + 2\dot{\tilde{\boldsymbol{\omega}}}\tilde{\boldsymbol{\omega}} + \tilde{\boldsymbol{\omega}}^3 \end{cases} \tag{43}$$

- hyper-jerk (jounce) field invariants

$$\begin{cases} \mathbf{a}_4 = \dddot{\mathbf{v}} + \tilde{\boldsymbol{\omega}}\ddot{\mathbf{v}} + \left(3\dot{\tilde{\boldsymbol{\omega}}} + \tilde{\boldsymbol{\omega}}^2\right)\dot{\mathbf{v}} + 3\ddot{\tilde{\boldsymbol{\omega}}}\mathbf{v} + 3\dot{\tilde{\boldsymbol{\omega}}}\tilde{\boldsymbol{\omega}}\mathbf{v} + 2\tilde{\boldsymbol{\omega}}\dot{\tilde{\boldsymbol{\omega}}}\mathbf{v} + \tilde{\boldsymbol{\omega}}^3\mathbf{v} \\ \boldsymbol{\Phi}_4 = \dddot{\tilde{\boldsymbol{\omega}}} + \tilde{\boldsymbol{\omega}}\ddot{\tilde{\boldsymbol{\omega}}} + \left(3\dot{\tilde{\boldsymbol{\omega}}} + \tilde{\boldsymbol{\omega}}^2\right)\dot{\tilde{\boldsymbol{\omega}}} + 3\ddot{\tilde{\boldsymbol{\omega}}}\tilde{\boldsymbol{\omega}} + 3\dot{\tilde{\boldsymbol{\omega}}}\tilde{\boldsymbol{\omega}}^2 + 2\tilde{\boldsymbol{\omega}}\dot{\tilde{\boldsymbol{\omega}}}\tilde{\boldsymbol{\omega}} + \tilde{\boldsymbol{\omega}}^4 \end{cases} \tag{44}$$

**Remark 5.** *The higher-order time derivative of spatial twist solve completely the problem of determining the field of the $n^{th}$ order acceleration of rigid motion.*

## 4.1 Higher-order acceleration center and vector invariants of rigid body motion

Equation (16) may be written as

$$\mathbf{a}_{\rho}^{[n]} - \phi_n \boldsymbol{\rho} = \mathbf{a}_{Q}^{[n]} - \boldsymbol{\Phi}_n \boldsymbol{\rho}_Q, n \in \mathbb{N}^*. \tag{45}$$

This shows us that the vector function

$$\mathbf{I}_n = \mathbf{a}_{\boldsymbol{\rho}}^{[n]} - \boldsymbol{\Phi}_n \boldsymbol{\rho}, n \in \mathbb{N}^* \tag{46}$$

has the same value in every point of the rigid body under the general spatial motion, at a given moment of time $t$. It represents a **vector invariant** of the n-th order acceleration field.

The invariant value of vector $\mathbf{I}_n$ is obtained for $\boldsymbol{\rho} = \mathbf{0}$ and it is the n-th order acceleration of the point of the rigid body that passes the origin of the fixed reference frame at a given moment of time: $\mathbf{I}_n = \mathbf{a}_0^{[n]} \stackrel{\text{def}}{=} \mathbf{a}_n$. Eq. (46) becomes:

$$\mathbf{a}_{\boldsymbol{\rho}}^{[n]} = \mathbf{a}_n + \phi_n \boldsymbol{\rho}. \tag{47}$$

Let be $\boldsymbol{\Phi}_n^*$ be the adjugate tensor of $\boldsymbol{\Phi}_n$ uniquely defined by:$\boldsymbol{\Phi}_n \boldsymbol{\Phi}_n^* = (\det \boldsymbol{\Phi}_n)\mathbf{I}$. From Eq. (46), results another invariant

$$\mathbf{J}_n = \boldsymbol{\Phi}_n^* \mathbf{a}_{\boldsymbol{\rho}}^{[n]} - (\det \boldsymbol{\Phi}_n)\boldsymbol{\rho}, n \in \mathbb{N}^*. \tag{48}$$

The value of this invariant is $\mathbf{J}_n = \boldsymbol{\Phi}_n^* \mathbf{a}_n$.

In the specific case when tensor $\boldsymbol{\Phi}_n$ is non-singular ($\det \boldsymbol{\Phi}_n \neq 0$), from (47) results the position vector having an imposed n-th order acceleration $\mathbf{a}^*$:

$$\boldsymbol{\rho}^* = \boldsymbol{\Phi}_n^{-1}(\mathbf{a}^* - \mathbf{a}_n), n \in \mathbb{N}^*. \tag{49}$$

In a particular case of the **n-th order acceleration center** $G_n$ (i.e. the point that have $\mathbf{a}^* = \mathbf{0}$) on obtain:

$$\boldsymbol{\rho}_{G_n} = -\boldsymbol{\Phi}_n^{-1} \mathbf{a}_n \tag{50}$$

Assuming that the tensor $\boldsymbol{\Phi}_n$ is non-singular, the previous relations lead to a new vector invariant that characterize the accelerations of n-th and m-th order ($n, m \in \mathbb{N}^*$):

$$\mathbf{K}_{m,n} = \mathbf{a}_{\boldsymbol{\rho}}^{[m]} - \boldsymbol{\Phi}_m \boldsymbol{\Phi}_n^{-1} \mathbf{a}_{\boldsymbol{\rho}}^{[n]}, m, n \in \mathbb{N}^*. \tag{51}$$

The value of this invariant is $\mathbf{K}_{m,n} = \mathbf{a}_m - \boldsymbol{\Phi}_m \boldsymbol{\Phi}_n^{-1} \mathbf{a}_n$.

The problem of the determination the adjugate tensor of the n-th acceleration tensor and the conditions in which these tensors are inversable is, as the author knows, still an open problem in theoretical kinematics field. We will propose a method based on the tensors algebra that will give a closed form, coordinate- free solution, dependent to the time derivative of spatial twist.

The vector field of the higher-order acceleration is a non-stationary vector field. Differential operator *div* and *curl* is expressed, taking into account Eq. (47), through the linear invariants of the tensor $\boldsymbol{\Phi}_n$, as below:

$$\begin{aligned} div\, \boldsymbol{a}_Q^{[n]} &= trace\, \boldsymbol{\Phi}_n \\ curl\, \boldsymbol{a}_Q^{[n]} &= 2vect\, \boldsymbol{\Phi}_n \end{aligned} \tag{52}$$

Let $\boldsymbol{\Phi} \in \mathbf{L}(\mathbf{V}_3, \mathbf{V}_3)$ a tensor and we note $\mathbf{t} = \text{vect}\boldsymbol{\Phi}$ and $\mathbf{S} = \text{sym}\boldsymbol{\Phi}$. The below theorem takes place.

**Theorem 4.** *The adjugate tensor and determinant of the tensor* $\boldsymbol{\Phi}$ *is*:

$$\begin{aligned} \boldsymbol{\Phi}^* &= \mathbf{S}^* - \widetilde{\mathbf{S}\mathbf{t}} + \mathbf{t} \otimes \mathbf{t} \\ \det\boldsymbol{\Phi} &= \det\mathbf{S} + \mathbf{t}\mathbf{S}\mathbf{t} \end{aligned} \tag{53}$$

Let $\boldsymbol{\Phi}_\text{n}$ the n-th order acceleration tensor, $\boldsymbol{\Phi}_\text{n} = \tilde{\mathbf{t}}_n + \mathbf{S}_n$.

The vectors $\mathbf{t}_n$ and the symmetric tensors $\mathbf{S}_n, n \in \mathbb{N}^*$ can be obtained with the below recurrence relation:

$$\begin{cases} \mathbf{t}_{\text{n}+1} = \dot{\mathbf{t}}_\text{n} + \dfrac{1}{2}\left[(\text{trace}\boldsymbol{\Phi}_\text{n})\mathbf{I} - \boldsymbol{\Phi}_\mathbf{n}^\text{T}\right]\boldsymbol{\omega} \\ \mathbf{t}_1 = \boldsymbol{\omega} \end{cases} \tag{54}$$

$$\begin{cases} \mathbf{S}_{\text{n}+1} = \dot{\mathbf{S}}_\text{n} + \text{sym}(\boldsymbol{\Phi}_\text{n}\tilde{\boldsymbol{\omega}}) \\ \mathbf{S}_1 = \mathbf{0} \end{cases} \tag{55}$$

It follows that:

- Velocity field: $\boldsymbol{\Phi}_1 = \tilde{\boldsymbol{\omega}}, \mathbf{t}_1 = \boldsymbol{\omega}, \mathbf{S}_1 = \mathbf{0}$

$$\begin{aligned} \boldsymbol{\Phi}_1^* &= \boldsymbol{\omega} \otimes \boldsymbol{\omega} \\ \det\boldsymbol{\Phi}_1 &= 0 \end{aligned} \tag{56}$$

$\boldsymbol{\Phi}_1$ is singular for any $\boldsymbol{\omega}$. In this case,

$$\begin{aligned} \text{div}\, \mathbf{a}_Q^{[1]} &= 0 \\ \text{curl}\, \mathbf{a}_Q^{[1]} &= 2\boldsymbol{\omega} \end{aligned} \tag{57}$$

- Acceleration field: $\boldsymbol{\Phi}_2 = \tilde{\boldsymbol{\omega}}^2 + \dot{\tilde{\boldsymbol{\omega}}}, \mathbf{t}_2 = \dot{\boldsymbol{\omega}}, \mathbf{S}_2 = \tilde{\boldsymbol{\omega}}^2$

$$\begin{aligned} \boldsymbol{\Phi}_2^* &= (\boldsymbol{\omega} \otimes \boldsymbol{\omega})^2 - \widetilde{\tilde{\boldsymbol{\omega}}^2\dot{\boldsymbol{\omega}}} + \dot{\boldsymbol{\omega}} \otimes \dot{\boldsymbol{\omega}} \\ \det\boldsymbol{\Phi}_2 &= -(\boldsymbol{\omega} \times \dot{\boldsymbol{\omega}})^2 \end{aligned} \tag{58}$$

$\boldsymbol{\Phi}_2$ is nonsingular if and only if $\boldsymbol{\omega} \times \dot{\boldsymbol{\omega}} \neq \mathbf{0}$. In this case

$$\boldsymbol{\Phi}_2^{-1} = \frac{\widetilde{\tilde{\boldsymbol{\omega}}^2\dot{\boldsymbol{\omega}}} - (\boldsymbol{\omega} \otimes \boldsymbol{\omega})^2 - \dot{\boldsymbol{\omega}} \otimes \dot{\boldsymbol{\omega}}}{(\boldsymbol{\omega} \times \dot{\boldsymbol{\omega}})^2} \tag{59}$$

$$\begin{aligned}\operatorname{div} \mathbf{a}_{\varrho}^{[2]} &= -2\omega^2\\ \operatorname{curl} \mathbf{a}_{\varrho}^{[2]} &= 2\dot{\omega}.\end{aligned} \tag{60}$$

- Jerk field: $\mathbf{\Phi}_3 = \ddot{\tilde{\omega}} + 2\dot{\tilde{\omega}}\tilde{\omega} + \tilde{\omega}\dot{\tilde{\omega}} + \tilde{\omega}^3, \mathbf{t}_3 = \ddot{\omega} + \frac{1}{2}\dot{\omega} \times \omega - \omega^2\omega, \mathbf{S}_3 = \frac{3}{2}\left[\tilde{\omega}\dot{\tilde{\omega}} + \dot{\tilde{\omega}}\tilde{\omega}\right],$

$$\mathbf{\Phi}_3^* = \frac{9}{4}\left[(\omega \otimes \dot{\omega})^2 + (\dot{\omega} \otimes \omega)^2 + (\omega \times \dot{\omega}) \otimes (\dot{\omega} \times \omega)\right] - \widetilde{\mathbf{S}_3\mathbf{t}_3} + \mathbf{t}_3 \otimes \mathbf{t}_3$$

$$\det\mathbf{\Phi}_3 = \frac{12(\mathbf{t}_3 \times \dot{\omega})(\omega \times \mathbf{t}_3) + 27\omega \cdot \dot{\omega}(\omega \times \dot{\omega})^2}{4}. \tag{61}$$

$\mathbf{\Phi}_3$ is nonsingular if and only if $4(\dot{\omega} \times \mathbf{t}_3)(\omega \times \mathbf{t}_3) \neq 9\omega \cdot \dot{\omega}(\omega \times \dot{\omega})^2$. In this case

$$\mathbf{\Phi}_3^{-1} == \frac{9\left[(\omega \otimes \dot{\omega})^2 + (\dot{\omega} \otimes \omega)^2 + (\omega \times \dot{\omega}) \otimes (\dot{\omega} \times \omega)\right] - 4\widetilde{\mathbf{S}_3\mathbf{t}_3} + 4\mathbf{t}_3 \otimes \mathbf{t}_3}{12(\mathbf{t}_3 \times \dot{\omega})(\omega \times \mathbf{t}_3) + 27\omega \cdot \dot{\omega}(\omega \times \dot{\omega})^2}$$

$$\begin{aligned}\operatorname{div} \mathbf{a}_{\varrho}^{[3]} &= -6\, \omega \cdot \dot{\omega}\\ \operatorname{curl} \mathbf{a}_{\varrho}^{[3]} &= 2\ddot{\omega} + \dot{\omega} \times \omega - 2\omega^2\omega.\end{aligned} \tag{62}$$

- Jounce field:

$$\begin{aligned}\mathbf{\Phi}_4 &= \dddot{\tilde{\omega}} + \tilde{\omega}\ddot{\tilde{\omega}} + \left(3\dot{\tilde{\omega}} + \tilde{\omega}^2\right)\dot{\tilde{\omega}} + 3\ddot{\tilde{\omega}}\tilde{\omega} + 3\dot{\tilde{\omega}}\tilde{\omega}^2 + 2\tilde{\omega}\dot{\tilde{\omega}}\tilde{\omega} + \tilde{\omega}^4\\ \mathbf{t}_4 &= \dddot{\omega} + \ddot{\omega} \times \omega - 2\omega^2\dot{\omega} - 4(\omega \cdot \dot{\omega})\omega\\ \mathbf{S}_4 &= 2\operatorname{sym}\left(2\ddot{\tilde{\omega}}\tilde{\omega} + \dot{\tilde{\omega}}\tilde{\omega}^2\right) + 3\dot{\tilde{\omega}}^2 + \tilde{\omega}^4\end{aligned} \tag{63}$$

$$\mathbf{S}_4^* = 2\operatorname{sym}\left[3\dot{\tilde{\omega}}(\omega \otimes \mathbf{w})\dot{\tilde{\omega}} - \alpha\omega \otimes \mathbf{w}\right] - \tilde{\omega}(\mathbf{w} \otimes \mathbf{w})\tilde{\omega} - 3\alpha\dot{\omega} \otimes \dot{\omega} + \alpha\omega^4\mathbf{I} \tag{64}$$

$$\mathbf{\Phi}_4^* = \mathbf{S}_4^* - \widetilde{\mathbf{S}_4\mathbf{t}_4} + \mathbf{t}_4 \otimes \mathbf{t}_4 \tag{65}$$

$$\det \mathbf{\Phi}_4 = \alpha\left[6(\omega \times \dot{\omega}) \cdot (\mathbf{w} \times \dot{\omega}) - (\omega \times \mathbf{w})^2 + 2\alpha\omega \cdot \mathbf{w} + 3\alpha\dot{\omega}^2 + \alpha^2\right] - 3(\dot{\omega}, \omega, \mathbf{w})^2 \tag{66}$$

In Eqs. (65) and (66), the following notation has been used:

$$\begin{aligned}\mathbf{w} &= \ddot{\omega} - \omega \times \dot{\omega} - \omega^2\omega\\ \alpha &= \omega^4 - 2\dot{\omega}^2 - 2\omega \cdot \dot{\omega}\end{aligned} \tag{67}$$

If

$$3(\dot{\omega}, \omega, \mathbf{w})^2 \neq \alpha\left[6(\omega \times \dot{\omega}) \cdot (\mathbf{w} \times \dot{\omega}) - (\omega \times \mathbf{w})^2 + 2\alpha\omega \cdot \mathbf{w} + 2\alpha\dot{\omega}^2 + \alpha^2\right] \tag{68}$$

then $\mathbf{\Phi}_4$ is inversible and

$$\mathbf{\Phi}_4{}^{-1} = \frac{\mathbf{\Phi}_4^*}{\det\mathbf{\Phi}_4} \tag{69}$$

In the hypothesis (68), there is jounce center, determined by

$$\begin{aligned} \boldsymbol{\rho}_{G_4} &= -\boldsymbol{\Phi}_4^{-1}\mathbf{a}_4 \\ \text{div}\,\mathbf{a}_\varrho^{[4]} &= -2\left(4\boldsymbol{\omega}\cdot\ddot{\boldsymbol{\omega}} + 3\dot{\boldsymbol{\omega}}^2 + \boldsymbol{\omega}^4\right) \\ \text{curl}\,\mathbf{a}_\varrho^{[4]} &= 2\dddot{\boldsymbol{\omega}} + 2\ddot{\boldsymbol{\omega}}\times\boldsymbol{\omega} - 4\boldsymbol{\omega}^2\dot{\boldsymbol{\omega}} - 8(\boldsymbol{\omega}\cdot\dot{\boldsymbol{\omega}})\boldsymbol{\omega} \end{aligned} \tag{70}$$

## 5. Dual algebra in rigid body kinematics

In this section, we will present some algebraic properties for dual numbers, dual vectors and dual tensors. More details can be found in [10–25].

### 5.1 Dual numbers

Let the set of real dual numbers to be denoted by

$$\underline{\mathbb{R}} = \left\{a + \varepsilon a_0 \middle| a, a_0 \in \mathbb{R}, \varepsilon^2 = 0, \varepsilon \neq 0\right\} \tag{71}$$

where $a = \text{Re}\,(\underline{a})$ is the real part of $\underline{a}$ and $a_0 = \text{Du}(\underline{a})$ the dual part. The sum and product between dual numbers generate a ring with zero divisors structure for $\underline{\mathbb{R}}$.

Any differentiable function $f : S \subset \mathbb{R} \to \mathbb{R}, f = f(a)$ can be completely defined on $\underline{S} \subset \underline{\mathbb{R}}$ such that:

$$f : \underline{S} \subset \underline{\mathbb{R}} \to \underline{\mathbb{R}}; f(\underline{a}) = f(a) + \varepsilon a_0 f'(a) \tag{72}$$

Based on the previous property, two of the most important functions have the following expressions: $\cos\underline{a} = \cos a - \varepsilon a_0 \sin a$; $\sin\underline{a} = \sin a + \varepsilon a_0 \cos a$;

### 5.2 Dual vectors

In the Euclidean space, the linear space of free vectors with dimension 3 will be denoted by $\mathbf{V}_3$. The ensemble of dual vectors is defined as:

$$\underline{\mathbf{V}_3} = \left\{\mathbf{a} + \varepsilon\mathbf{a}_0; \mathbf{a}, \mathbf{a}_0 \in \mathbf{V}_3, \varepsilon^2 = 0, \varepsilon \neq 0\right\} \tag{73}$$

where $\mathbf{a} = \text{Re}\,(\underline{\mathbf{a}})$ is the real part of $\underline{\mathbf{a}}$ and $\mathbf{a}_0 = \text{Du}(\underline{\mathbf{a}})$ the dual part. For any three dual vectors $\underline{\mathbf{a}}, \underline{\mathbf{b}}, \underline{\mathbf{c}}$, the following notations will be used for the basic products: $\underline{\mathbf{a}}\cdot\underline{\mathbf{b}}$—scalar product, $\underline{\mathbf{a}}\times\underline{\mathbf{b}}$—cross product and $\langle\underline{\mathbf{a}}, \underline{\mathbf{b}}, \underline{\mathbf{c}}\rangle = \underline{\mathbf{a}}\cdot(\underline{\mathbf{b}}\times\underline{\mathbf{c}})$—triple scalar product. Regarding algebraic structure, $(\underline{\mathbf{V}}_3, +, \cdot_{\underline{\mathbb{R}}})$ is a free $\underline{\mathbb{R}}$-module [13, 14].

The magnitude of $|\underline{\mathbf{a}}|$, denoted by $|\underline{\mathbf{a}}|$, is the dual number computed from

$$\underline{\mathbf{a}} = \begin{cases} \|\mathbf{a}\| + \varepsilon\dfrac{\mathbf{a}_0\cdot\mathbf{a}}{\|\mathbf{a}\|}, \ \text{Re}\,(\underline{\mathbf{a}}) \neq \mathbf{0} \\ \varepsilon\|\mathbf{a}_0\|, \ \text{Re}\,(\underline{\mathbf{a}}) = \mathbf{0} \end{cases} \tag{74}$$

where $\|.\|$ is the Euclidean norm. For any dual vector $\underline{\mathbf{a}} \in \underline{\mathbf{V}}_3$, if $|\underline{\mathbf{a}}| = 1$ then $\underline{\mathbf{a}}$ is called unit dual vector.

### 5.3 Dual tensors

An $\underline{\mathbb{R}}$-linear application of $\underline{\mathbf{V}}_3$ into $\underline{\mathbf{V}}_3$ is called an Euclidean dual tensor:

$$\underline{\mathbf{T}}(\underline{\lambda}_1\underline{\mathbf{v}}_1 + \underline{\lambda}_2\underline{\mathbf{v}}_2) = \underline{\lambda}_1\underline{\mathbf{T}}(\underline{\mathbf{v}}_1) + \underline{\lambda}_2\underline{\mathbf{T}}(\underline{\mathbf{v}}_2), \forall\underline{\lambda}_1, \underline{\lambda}_2 \in \underline{\mathbb{R}}, \forall\underline{\mathbf{v}}_1, \underline{\mathbf{v}}_2 \in \underline{\mathbf{V}}_3 \tag{75}$$

Let $\mathbf{L}(\underline{\mathbf{V}}_3, \underline{\mathbf{V}}_3)$ be the set of dual tensors, then any dual tensor $\underline{\mathbf{T}} \in \mathbf{L}(\underline{\mathbf{V}}_3, \underline{\mathbf{V}}_3)$ can be decomposed as $\underline{\mathbf{T}} = \mathbf{T} + \varepsilon\mathbf{T}_0$, where $\mathbf{T}, \mathbf{T}_0 \in \mathbf{L}(\mathbf{V}_3, \mathbf{V}_3)$ are real tensors. Also, the dual transposed tensor, denoted by $\underline{\mathbf{T}}^{\mathrm{T}}$, is defined by

$$\underline{\mathbf{v}}_1 \cdot (\underline{\mathbf{T}}\underline{\mathbf{v}}_1) = \underline{\mathbf{v}}_2 \cdot \left(\underline{\mathbf{T}}^{\mathrm{T}}\underline{\mathbf{v}}_1\right), \forall\underline{\mathbf{v}}_1, \underline{\mathbf{v}}_2 \in \underline{\mathbf{V}}_3 \tag{76}$$

while $\forall\underline{\mathbf{v}}_1, \underline{\mathbf{v}}_2, \underline{\mathbf{v}}_3 \in \underline{\mathbf{V}}_3,\ \mathrm{Re}\left\langle\underline{\mathbf{v}}_1, \underline{\mathbf{v}}_2, \underline{\mathbf{v}}_3\right\rangle \neq 0$ the determinant is

$$\left\langle\underline{\mathbf{T}}\underline{\mathbf{v}}_1, \underline{\mathbf{T}}\underline{\mathbf{v}}_2, \underline{\mathbf{T}}\underline{\mathbf{v}}_3\right\rangle = \det\underline{\mathbf{T}}\left\langle\underline{\mathbf{v}}_1, \underline{\mathbf{v}}_2, \underline{\mathbf{v}}_3\right\rangle. \tag{77}$$

Orthogonal dual tensor maps are a powerful instrument in the study of the rigid motion with respect to an inertial and non-inertial reference frames.

Let the orthogonal dual tensor set be denoted by:

$$\underline{\mathbb{SO}}_3 = \left\{\underline{\mathbf{R}} \in \mathbf{L}(\underline{\mathbf{V}}_3, \underline{\mathbf{V}}_3) \middle| \underline{\mathbf{R}}\underline{\mathbf{R}}^{\mathrm{T}} = \underline{\mathbf{I}}, \det\underline{\mathbf{R}} = 1\right\} \tag{78}$$

where $\underline{\mathbb{SO}}_3$ is the set of special orthogonal dual tensors and $\underline{\mathbf{I}}$ is the unit orthogonal dual tensor.

**Theorem 5 (Structure Theorem).** *For any $\underline{\boldsymbol{R}} \in \underline{\mathbb{SO}}_3$ a unique decomposition is viable*

$$\underline{\mathbf{R}} = (\mathbf{I} + \varepsilon\tilde{\boldsymbol{\rho}})\mathbf{R} \tag{79}$$

*where $\boldsymbol{R} \in \mathbb{SO}_3$ and $\boldsymbol{\rho} \in V_3$ are called* ***structural invariants.***

Taking into account the Lie group structure of $\underline{\mathbb{SO}}_3$ and the result presented in previous theorem, it can be concluded that any orthogonal dual tensor $\underline{\mathbf{R}} \in \underline{\mathbb{SO}}_3$ can be used globally parameterize displacements of rigid bodies.

**Theorem 6 (Representation Theorem).** *For any orthogonal dual tensor $\underline{\mathbf{R}}$ defined as in Eq. (79), a dual number $\underline{\alpha} = \alpha + \varepsilon d$ and a dual unit vector $\underline{\mathbf{u}} = \mathbf{u} + \varepsilon\mathbf{u}_0$ can be computed to have the following equation*, [13–15]:

$$\underline{\mathbf{R}}(\underline{\alpha}, \underline{\mathbf{u}}) = \underline{\mathbf{I}} + \sin\underline{\alpha}\underline{\tilde{\mathbf{u}}} + (1 - \cos\underline{\alpha})\underline{\tilde{\mathbf{u}}}^2 = \exp\left(\underline{\alpha}\underline{\tilde{\mathbf{u}}}\right) \tag{80}$$

The parameters $\underline{\alpha}$ and $\underline{\mathbf{u}}$ are called the **natural invariants** of $\underline{\mathbf{R}}$. The unit dual vector $\underline{\mathbf{u}}$ gives the Plücker representation of the Mozzi-Chasles axis [13, 14], while the dual angle $\underline{\alpha} = \alpha + \varepsilon d$ contains the rotation angle $\alpha$ and the translated distance d.

The Lie algebra of the Lie group $\underline{\mathbb{SO}}_3$ is the skew-symmetric dual tensor set denoted by $\underline{\mathfrak{so}}_3 = \left\{\underline{\tilde{\boldsymbol{\alpha}}} \in \mathbf{L}(\underline{\mathbf{V}}_3, \underline{\mathbf{V}}_3) \middle| \underline{\tilde{\boldsymbol{\alpha}}} = -\underline{\tilde{\boldsymbol{\alpha}}}^{\mathrm{T}}\right\}$, where the internal mapping is $\langle\underline{\tilde{\boldsymbol{\alpha}}}_1, \underline{\tilde{\boldsymbol{\alpha}}}_2\rangle = \widetilde{\underline{\tilde{\boldsymbol{\alpha}}}_1\underline{\boldsymbol{\alpha}}_2}$.

The link between the Lie algebra $\underline{\mathfrak{so}}_3$, the Lie group $\underline{\mathbb{SO}}_3$, and the exponential map is given by the following.

**Theorem 7.** *The mapping*

$$exp\colon \underline{\mathfrak{so}}_3 \to \underline{\mathbb{SO}}_3,\ exp\left(\underline{\tilde{\boldsymbol{\alpha}}}\right) = e^{\underline{\tilde{\boldsymbol{\alpha}}}} = \sum_{k=0}^{\infty}\frac{\underline{\tilde{\boldsymbol{\alpha}}}^k}{k!} \tag{81}$$

is well defined and onto.

$$\log\colon \underline{\mathbb{SO}}_3 \to \underline{\mathfrak{so}}_3,\ \log\underline{\mathbf{R}} = \left\{\underline{\tilde{\boldsymbol{\psi}}} \in \underline{\mathfrak{so}}_3 \middle| \exp\left(\underline{\tilde{\boldsymbol{\psi}}}\right) = \underline{\mathbf{R}}\right\} \tag{82}$$

and is the inverse of Eq. (81).

Based on Theorems 6 and 7, for any orthogonal dual tensor $\underline{\mathbf{R}}$, a dual vector $\underline{\boldsymbol{\psi}} = \underline{\alpha}\,\underline{\mathbf{u}} = \boldsymbol{\psi} + \varepsilon\boldsymbol{\psi}_0$ can be computed and represents the **screw dual vector**, which embeds the screw axis and screw parameters.

The form of $\underline{\boldsymbol{\psi}}$ implies that $\tilde{\underline{\boldsymbol{\psi}}} \in \log \underline{\mathbf{R}}$. The types of rigid displacements that can be parameterized by $\underline{\boldsymbol{\psi}}$ are:

- roto-translation if $\boldsymbol{\psi} \neq 0, \boldsymbol{\psi}_0 \neq 0$ and $\boldsymbol{\psi} \cdot \boldsymbol{\psi}_0 \neq 0 \Leftrightarrow$ if $\left|\underline{\boldsymbol{\psi}}\right| \in \underline{\mathbb{R}}$ and $\left|\underline{\boldsymbol{\psi}}\right| \notin \{\varepsilon\mathbb{R}\}$;
- pure translation if if $\boldsymbol{\psi} = 0$ and $\boldsymbol{\psi}_0 \neq 0 \Leftrightarrow$ if $\left|\underline{\boldsymbol{\psi}}\right| \in \underline{\varepsilon\mathbb{R}}$;
- pure rotation if $\boldsymbol{\psi} \neq 0$ and $\boldsymbol{\psi} \cdot \boldsymbol{\psi}_0 = 0 \Leftrightarrow$ if $\left|\underline{\boldsymbol{\psi}}\right| \in \mathbb{R}$.

**Theorem 8.** *The natural invariants* $\underline{\alpha} = \alpha + \varepsilon d$, $\underline{\mathbf{u}} = \mathbf{u} + \varepsilon\mathbf{u}_0$ *can be used to directly recover the structural invariants* $\mathbf{R}$ *and* $\boldsymbol{\rho}$ *from Eq. (79):*

$$\begin{aligned} \mathbf{R} &= \mathbf{I} + \sin\alpha\tilde{\mathbf{u}} + (1 - \cos\alpha)\tilde{\mathbf{u}}^2 \\ \boldsymbol{\rho} &= d\mathbf{u} + \sin\alpha\mathbf{u}_0 + (1 - \cos\alpha)\mathbf{u} \times \mathbf{u}_0 \end{aligned} \tag{83}$$

**Theorem 9 (isomorphism theorem).** *The special Euclidean group* $(\mathbb{SE}_3, \cdot)$ *and* $(\underline{\mathbb{SO}}_3, \cdot)$ *are connected via the isomorphism of the Lie groups*

$$\begin{aligned} &\Phi : \mathbb{SE}_3 \to \underline{\mathbb{SO}}_3 \\ &\Phi(g) = (\mathbf{I} + \varepsilon\tilde{\boldsymbol{\rho}})\mathbf{R} \end{aligned} \tag{84}$$

*where* $g = \begin{bmatrix} \mathbf{R} & \boldsymbol{\rho} \\ \mathbf{0} & 1 \end{bmatrix}$, $\mathbf{R} \in \underline{\mathbb{SO}}_3$, $\boldsymbol{\rho} \in \mathbf{V}_3$.

**Remark 6.** *The inverse of* $\boldsymbol{\Phi}$ is

$$\boldsymbol{\Phi}^{-1} : \underline{\mathbb{SO}}_3 \leftrightarrow \mathbb{SE}_3; \boldsymbol{\Phi}^{-1}(\underline{\mathbf{R}}) = \begin{bmatrix} \mathbf{R} & \boldsymbol{\rho} \\ \mathbf{0} & 1 \end{bmatrix} \tag{85}$$

*where* $\mathbf{R} = \mathrm{Re}\,(\underline{\mathbf{R}}), \boldsymbol{\rho} = \mathrm{vect}\left(\mathrm{Du}(\underline{\mathbf{R}}) \cdot \mathbf{R}^T\right)$.

## 6. Higher-order kinematics in dual Lie algebra

Being the rigid body motion given by the following parametric equation in a given reference frame:

$$\begin{cases} \boldsymbol{\rho} = \boldsymbol{\rho}(t) \in \mathbf{V}_3 \\ \mathbf{R} = \mathbf{R}(t) \in \mathbb{SO}_3 \end{cases} \tag{86}$$

with $t \in \mathbf{I} \subseteq \mathbb{R}$ is time variable.

The dual orthogonal tensor that describes the rigid body motion is [13, 24]:

$$\underline{\mathbf{R}} = (\mathbf{I} + \varepsilon\tilde{\boldsymbol{\rho}})\mathbf{R} \tag{87}$$

In relation (87), the skew symmetric tensor associated to the vector $\boldsymbol{\rho}$ is denoted by $\tilde{\boldsymbol{\rho}}$.

It can be easily demonstrated [14, 15], that:

$$\underline{\mathbf{R}}\underline{\mathbf{R}}^{\mathrm{T}} = \underline{\mathbf{I}}$$
$$\det\underline{\mathbf{R}} = 1$$
$$\underline{\mathbf{a}} \cdot \underline{\mathbf{b}} = (\underline{\mathbf{R}}\underline{\mathbf{a}}) \cdot (\underline{\mathbf{R}}\underline{\mathbf{b}}), \forall \underline{\mathbf{a}}, \underline{\mathbf{b}} \in \underline{\mathbf{V}}_3$$
$$\underline{\mathbf{R}}(\underline{\mathbf{a}} \times \underline{\mathbf{b}}) = \underline{\mathbf{R}}\,(\underline{\mathbf{a}}) \times \underline{\mathbf{R}}\,(\underline{\mathbf{b}}), \forall \underline{\mathbf{a}}, \underline{\mathbf{b}} \in \underline{\mathbf{V}}_3 \tag{88}$$

The tensor **R** transports the dual vectors from the body frame in the space frame with the conservation of the dual angles and the relative orientation of lines that corresponds to the dual vectors $\underline{\mathbf{a}}$ and $\underline{\mathbf{b}}$.

The dual angular velocity for the rigid body motion (86) is given by (87):

$$\underline{\boldsymbol{\omega}} = \mathrm{vect}\dot{\underline{\mathbf{R}}}\underline{\mathbf{R}}^{\mathrm{T}} \tag{89}$$

It can be demonstrated that:

$$\underline{\boldsymbol{\omega}} = \boldsymbol{\omega} + \varepsilon\mathbf{v} \tag{90}$$

where

$$\boldsymbol{\omega} = \mathrm{vect}\dot{\mathbf{R}}\mathbf{R}^{\mathrm{T}} \tag{91}$$

is the instantaneous angular velocity of the rigid body and

$$\mathbf{v} = \dot{\boldsymbol{\rho}} - \boldsymbol{\omega} \times \boldsymbol{\rho} \tag{92}$$

is the linear velocity of a point of the rigid body that coincides with the origin of the reference frame at that given moment.

The dual angular velocity $\underline{\boldsymbol{\omega}}$ completely characterizes the distribution of the velocity field of the rigid body. The pair $(\boldsymbol{\omega}, \mathbf{v})$ is called "the twist of the rigid body motion" [13, 14].

Being:

$$\underline{\mathbf{V}}_{\rho} = \boldsymbol{\omega} + \varepsilon\mathbf{v}_{\rho} \tag{93}$$

the dual velocity for a point localized in the reference frame by the position vector $\boldsymbol{\rho}$.

In (93), $\boldsymbol{\omega}$ is the instantaneous angular velocity of the rigid body and $\mathbf{v}_{\rho}$ is the linear velocity of the point. Using the next equation,

$$e^{\varepsilon\tilde{\boldsymbol{\rho}}} = \mathbf{I} + \varepsilon\tilde{\boldsymbol{\rho}} \tag{94}$$

from (90), (92)–(94), results:

$$e^{\varepsilon\tilde{\boldsymbol{\rho}}}\underline{\mathbf{V}}_{\rho} = \underline{\boldsymbol{\omega}} \tag{95}$$

Consequently, $e^{\varepsilon\tilde{\mathbf{r}}}\underline{\mathbf{V}}_{r}$ is an invariant having the same value for any **r**.

Writing this invariant in two different points of the rigid body (noted with P and Q ), results that:

$$e^{\varepsilon\tilde{\mathbf{r}}_P}\underline{\mathbf{V}}_{P} = e^{\varepsilon\tilde{\mathbf{r}}_Q}\underline{\mathbf{V}}_{Q} \tag{96}$$

From (97), results:

$$\underline{\mathbf{V}}_P = e^{\varepsilon\widetilde{PQ}}\underline{\mathbf{V}}_Q \tag{97}$$

with $PQ = \boldsymbol{\rho}_Q - \boldsymbol{\rho}_P$.

Relation (97) is true for any P and Q.

Analogue with Eq. (95), the following invariants take place:

$$\begin{aligned} e^{\varepsilon\tilde{\boldsymbol{\rho}}}\underline{\mathbf{A}}_\rho &= \underline{\dot{\boldsymbol{\omega}}}, \forall \boldsymbol{\rho} \in \mathbf{V}_3 \\ e^{\varepsilon\tilde{\boldsymbol{\rho}}}\underline{\mathbf{J}}_\rho &= \underline{\ddot{\boldsymbol{\omega}}}, \forall \boldsymbol{\rho} \in \mathbf{V}_3 \\ e^{\varepsilon\tilde{\boldsymbol{\rho}}}\underline{\mathbf{H}}_\rho &= \underline{\dddot{\boldsymbol{\omega}}}, \forall \boldsymbol{\rho} \in \mathbf{V}_3 \end{aligned} \tag{98}$$

where we denoted

$$\begin{aligned} \underline{\mathbf{A}}_\rho &= \dot{\boldsymbol{\omega}} + \varepsilon \mathbf{A}_\rho \\ \underline{\mathbf{J}}_\rho &= \ddot{\boldsymbol{\omega}} + \varepsilon \mathbf{J}_\rho \\ \underline{\mathbf{H}}_\rho &= \dddot{\boldsymbol{\omega}} + \varepsilon \mathbf{H}_\rho \end{aligned} \tag{99}$$

with $\mathbf{A}_\rho, \mathbf{J}_\rho, \mathbf{H}_\rho$ the reduced acceleration, reduced jerk, respectively the reduced hyper-jerk (jounce), in a point given by the position vector $\boldsymbol{\rho}$:

$$\begin{aligned} \mathbf{A}_\rho &= \mathbf{a}_\rho - \boldsymbol{\omega} \times \mathbf{v}_\rho \\ \mathbf{J}_\rho &= \mathbf{j}_\rho - \boldsymbol{\omega} \times \mathbf{a}_\rho - 2\dot{\boldsymbol{\omega}} \times \mathbf{v}_\rho \\ \mathbf{H}_\rho &= \mathbf{h}_\rho - \boldsymbol{\omega} \times \mathbf{j}_\rho - 3\dot{\boldsymbol{\omega}} \times \mathbf{a}_\rho - 3\ddot{\boldsymbol{\omega}} \times \mathbf{v}_\rho \end{aligned} \tag{100}$$

In (100), $\mathbf{a}_\rho$, $\mathbf{j}_\rho$ and $\mathbf{h}_\rho$ are, respectively, the acceleration, the jerk, and the hyper-jerk (jounce), in a point given by the position vector $\boldsymbol{\rho}$.

Analogue with Eq. (97) the following equations take place:

$$\begin{aligned} \underline{\mathbf{A}}_P &= e^{\varepsilon\widetilde{PQ}}\underline{\mathbf{A}}_Q \\ \underline{\mathbf{J}}_P &= e^{\varepsilon\widetilde{PQ}}\underline{\mathbf{J}}_Q \ , \\ \underline{\mathbf{H}}_P &= e^{\varepsilon\widetilde{PQ}}\underline{\mathbf{H}}_Q \end{aligned} \tag{101}$$

The lines corresponding to the dual vectors $\underline{\dot{\boldsymbol{\omega}}}, \underline{\ddot{\boldsymbol{\omega}}}, \underline{\dddot{\boldsymbol{\omega}}}$ represent the loci, where the vectors $\mathbf{A}_\rho, \mathbf{J}_\rho, \mathbf{H}_\rho$ have the minimum module value. Supplementary,

$$\begin{aligned} \min_{\rho \in V_3} \left\|\mathbf{A}_\rho\right\| &= \left|\mathrm{Du}\left|\underline{\dot{\boldsymbol{\omega}}}\right|\right| \\ \min_{\rho \in V_3} \left\|\mathbf{J}_\rho\right\| &= \left|\mathrm{Du}\left|\underline{\ddot{\boldsymbol{\omega}}}\right|\right| \\ \min_{\rho \in V_3} \left\|\mathbf{H}_\rho\right\| &= \left|\mathrm{Du}\left|\underline{\dddot{\boldsymbol{\omega}}}\right|\right| \end{aligned} \tag{102}$$

Interesting is the fact that for the plane motion $\min \left\|\mathbf{A}_\rho\right\| = \min \left\|\mathbf{J}_\rho\right\| = \min \left\|\mathbf{H}_\rho\right\| = 0$ because $\mathrm{Du}\left|\underline{\dot{\boldsymbol{\omega}}}\right| = \mathrm{Du}\left|\underline{\ddot{\boldsymbol{\omega}}}\right| = \mathrm{Du}\left|\underline{\dddot{\boldsymbol{\omega}}}\right| = 0$

All properties are extended for higher-order accelerations. The vector $\underline{\boldsymbol{\omega}}^{(n)} = \frac{d^n \underline{\boldsymbol{\omega}}}{dt^n}, n \in \mathbb{N}$ describes completely the helicoidally field of the $n$ order reduced accelerations, for $n \in \mathbb{N}$:

$$e^{\varepsilon\tilde{\rho}}\underline{\mathbf{A}}_{\rho}^{[n]} = \underline{\boldsymbol{\omega}}^{(n)} \tag{103}$$

In Eq. (103) $\underline{\mathbf{A}}_{\rho}^{[n]}$ denote the $n^{th}$ order of the dual reduced acceleration in a given point by the position vector $\boldsymbol{\rho}$.

It follows that the dual part of the $n^{th}$ order differentiation of $\underline{\boldsymbol{\omega}}^{(n)}$

$$\underline{\boldsymbol{\omega}}^{(n)} = \boldsymbol{\omega}^{(n)} + \varepsilon\mathbf{v}^{(n)} \tag{104}$$

is the $n^{th}$ order reduced acceleration of that point of the rigid body that at the given time pass by the origin of the reference frame.

From equation

$$\mathbf{v} = \dot{\boldsymbol{\rho}} - \boldsymbol{\omega} \times \boldsymbol{\rho} \tag{105}$$

it follows that

$$\mathbf{v}^{(n)} = \boldsymbol{\rho}^{(n+1)} - \sum_{k=0}^{n} C_n^k \boldsymbol{\omega}^{(k)} \times \boldsymbol{\rho}^{(n-k)}, n \in \mathbb{N} \tag{106}$$

with the following notations

$$\mathbf{a}_{\rho}^{[n]} \triangleq \boldsymbol{\rho}^{(n+1)}, n \in \mathbb{N} \tag{107}$$

for the $n \in \mathbb{N}$ order acceleration of the point given by the position vector $\boldsymbol{\rho}$ and

$$\mathbf{A}_{\rho}^{[n]} \triangleq \mathbf{a}_{\rho}^{[n]} - \sum_{k=0}^{n-1} C_n^k \boldsymbol{\omega}^{(n-k)} \mathbf{a}_{\rho}^{[k]} \tag{108}$$

for the $n^{th}$ order reduced acceleration of the same point the equation:

$$\mathbf{A}_{\rho}^{[n]} = \mathbf{v}^{(n)} + \boldsymbol{\omega}^{(n)} \times \boldsymbol{\rho} \tag{109}$$

which proves the character of the helicoidally field of the $n^{th}$ order reduced accelerations field.

For $\boldsymbol{\rho} = 0$, the relations between the $n^{th}$ order reduced acceleration and the n order acceleration from point O, the origin of the reference frame, are written

$$\mathbf{A}_0^{[n]} = \mathbf{v}^{(n)} = \mathbf{a}_n - \sum_{k=1}^{n-1} C_n^k \boldsymbol{\omega}^{(n-k)} \mathbf{a}_k, n \in \mathbb{N} \tag{110}$$

The invert of previous equation is written:

$$\mathbf{a}_n = \mathbf{P}_n(\mathbf{v}), n \in \mathbb{N} \tag{111}$$

where $\mathbf{P}_n$ is the polynomial with the coefficients in the ring of the second order Euclidean tensors and the polynomials $\mathbf{P}_n[\mathbf{D}]$ follow the recurrence equation:

$$\begin{cases} \mathbf{P}_{n+1} = \mathbf{D}\mathbf{P}_n + \mathbf{P}_n(\tilde{\boldsymbol{\omega}}) \\ \mathbf{P}_0 = \mathbf{I} \end{cases} \tag{112}$$

it follows successively

$$
\begin{aligned}
\mathbf{P}_1 &= \tilde{\boldsymbol{\omega}} \\
\mathbf{P}_2 &= \mathbf{D} + \tilde{\boldsymbol{\omega}} \\
\mathbf{P}_3 &= \mathbf{D}^2 + \tilde{\boldsymbol{\omega}}\mathbf{D} + 2\dot{\tilde{\boldsymbol{\omega}}} + \tilde{\boldsymbol{\omega}}^2 \\
\mathbf{P}_4 &= \mathbf{D}^3 + \tilde{\boldsymbol{\omega}}\mathbf{D}^2 + \left(3\dot{\tilde{\boldsymbol{\omega}}} + \tilde{\boldsymbol{\omega}}^2\right)\mathbf{D} + 3\ddot{\tilde{\boldsymbol{\omega}}} + 2\tilde{\boldsymbol{\omega}}\dot{\tilde{\boldsymbol{\omega}}} + 3\dot{\tilde{\boldsymbol{\omega}}}\tilde{\boldsymbol{\omega}} + \tilde{\boldsymbol{\omega}}^3
\end{aligned}
\tag{113}
$$

If we denote $\mathbf{T} = (\mathbf{v}, \tilde{\boldsymbol{\omega}})$ and by $\boldsymbol{\Psi}_n = (\mathbf{a}_n, \boldsymbol{\Phi}_n)$, $n \in \mathbb{N}^*$, for the case of the velocities, accelerations, jerks and jounces, on obtain (**Figure 1**):

$$
\begin{bmatrix} \mathbf{T} \\ \dot{\mathbf{T}} \\ \ddot{\mathbf{T}} \\ \ddot{\mathbf{T}} \end{bmatrix} = \begin{bmatrix} \mathbf{I} & \mathbf{0} & \mathbf{0} & \mathbf{0} \\ -\tilde{\boldsymbol{\omega}} & \mathbf{I} & \mathbf{0} & \mathbf{0} \\ -2\dot{\tilde{\boldsymbol{\omega}}} & -\tilde{\boldsymbol{\omega}} & \mathbf{I} & \mathbf{0} \\ -3\ddot{\tilde{\boldsymbol{\omega}}} & -3\dot{\tilde{\boldsymbol{\omega}}} & -\tilde{\boldsymbol{\omega}} & \mathbf{I} \end{bmatrix} \begin{bmatrix} \boldsymbol{\Psi}_1 \\ \boldsymbol{\Psi}_2 \\ \boldsymbol{\Psi}_3 \\ \boldsymbol{\Psi}_4 \end{bmatrix} \tag{114}
$$

$$
\begin{bmatrix} \boldsymbol{\Psi}_1 \\ \boldsymbol{\Psi}_2 \\ \boldsymbol{\Psi}_3 \\ \boldsymbol{\Psi}_4 \end{bmatrix} = \begin{bmatrix} \mathbf{I} & \mathbf{0} & \mathbf{0} & \mathbf{0} \\ \tilde{\boldsymbol{\omega}} & \mathbf{I} & \mathbf{0} & \mathbf{0} \\ 2\dot{\tilde{\boldsymbol{\omega}}} + \tilde{\boldsymbol{\omega}}^2 & \tilde{\boldsymbol{\omega}} & \mathbf{I} & \mathbf{0} \\ 3\ddot{\tilde{\boldsymbol{\omega}}} + 2\tilde{\boldsymbol{\omega}}\dot{\tilde{\boldsymbol{\omega}}} + 3\dot{\tilde{\boldsymbol{\omega}}}\tilde{\boldsymbol{\omega}} + \tilde{\boldsymbol{\omega}}^3 & 3\dot{\tilde{\boldsymbol{\omega}}} + \tilde{\boldsymbol{\omega}}^2 & \tilde{\boldsymbol{\omega}} & \mathbf{I} \end{bmatrix} \begin{bmatrix} \mathbf{T} \\ \dot{\mathbf{T}} \\ \ddot{\mathbf{T}} \\ \dddot{\mathbf{T}} \end{bmatrix} \tag{115}
$$

$$
\mathbf{T}^{(\mathrm{n}-1)} = \boldsymbol{\Psi}_{\mathrm{n}} - \sum_{\mathrm{k}=1}^{\mathrm{n}-1} \mathbf{C}_{\mathrm{n}-1}^{\mathrm{k}} \tilde{\boldsymbol{\omega}}^{(\mathrm{n}-1-\mathrm{k})} \boldsymbol{\Psi}_{\mathrm{k}}
$$

$$
\boldsymbol{\Psi}_{\mathrm{n}} = \mathbf{P}_{\mathrm{n}}\mathbf{T}, \mathrm{n} \in \mathbb{N}^* \tag{116}
$$

**Theorem 10.** *The $n^{th}$ order accelerations field of a rigid body in a general motion is uniquely determined by the $k^{th}$ order time derivative of a dual twist $\underline{\boldsymbol{\omega}}$, $k = \overline{0, n-1}$.*

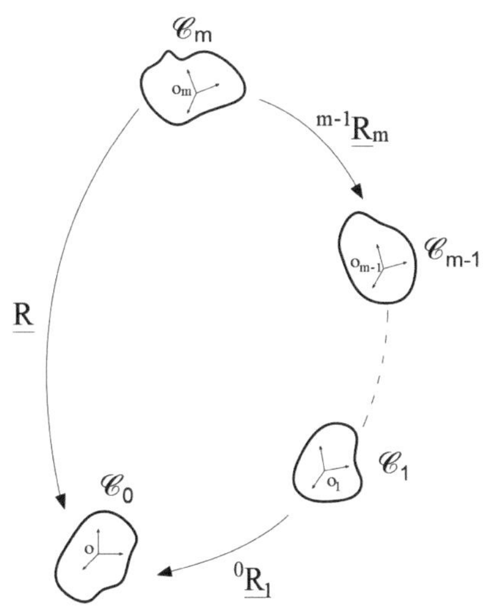


**Figure 1.**
*Higher-order time derivative of dual twist.*

## 7. Higher-order kinematics of spatial chain using dual Lie algebra

Consider a spatial kinematic chain of the bodies $C_k, k = \overline{0, m}$ where the relative motion of the rigid body $C_k$ with respect to $C_{k-1}$ is given by the proper orthogonal tensor ${}^{k-1}\underline{\mathbf{R}}_k \in \underline{\mathbf{SO}}_3^{\mathbb{R}}$. The relative motion properties of the body $C_m$ with respect to $C_0$ are described by the orthogonal dual tensor (**Figure 2**):

$$\underline{\mathbf{R}} = {}^0\underline{\mathbf{R}}_1 \cdot {}^1\underline{\mathbf{R}}_2 \ldots {}^{m-1}\underline{\mathbf{R}}_m \tag{117}$$

Instantaneous dual angular velocity (dual twist) of the rigid body in relation to the reference frame it will be given by the equation

$$_0\underline{\boldsymbol{\omega}}_m = \text{vect}\dot{\underline{\mathbf{R}}}\underline{\mathbf{R}}^T \tag{118}$$

It follows from (110) and (111) that:

$$_0\underline{\boldsymbol{\omega}}_m = \underline{\boldsymbol{\Omega}}_1 + {}^0\mathbf{R}_1\underline{\boldsymbol{\Omega}}_2 + \ldots + {}^0\mathbf{R}_1 \cdot {}^1\mathbf{R}_2 \ldots {}^{m-2}\mathbf{R}_{m-1}\underline{\boldsymbol{\Omega}}_m \tag{119}$$

where

$$\underline{\boldsymbol{\Omega}}_k = \text{vect}\,{}^{k-1}\dot{\underline{\mathbf{R}}}_k\,{}^{k-1}\underline{\mathbf{R}}_k^T \tag{120}$$

Using the denotation

$$\underline{\boldsymbol{\omega}}_k = {}^0\underline{\mathbf{R}}_1 \cdot {}^1\underline{\mathbf{R}}_2 \ldots {}^{k-2}\underline{\mathbf{R}}_{k-1}\underline{\boldsymbol{\Omega}}_k \tag{121}$$

Eq. (118) will be written

$$_0\underline{\boldsymbol{\omega}}_m = \underline{\boldsymbol{\omega}}_1 + \underline{\boldsymbol{\omega}}_2 + \ldots + \underline{\boldsymbol{\omega}}_m \tag{122}$$

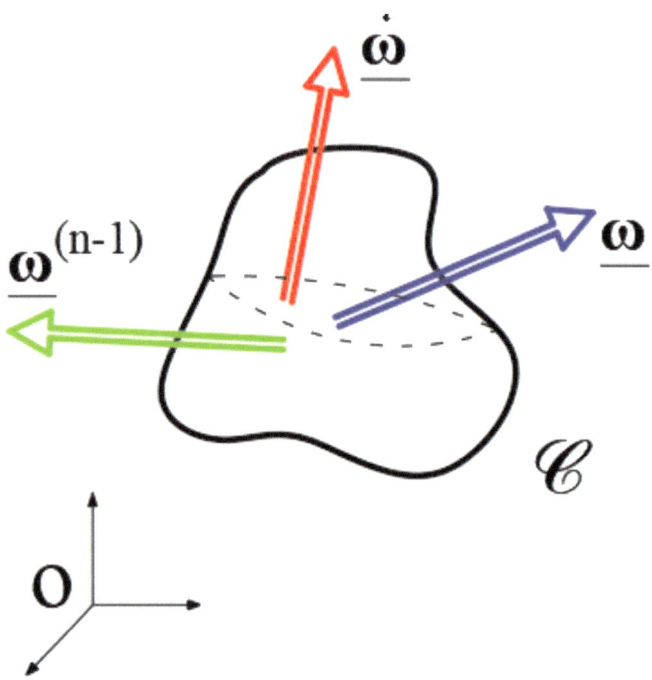


**Figure 2.**
*Orthogonal dual tensors of relative rigid body motion.*

where $\underline{\boldsymbol{\omega}}_k$ is the dual twist of the relative motion of the body $C_k$ in relation to the body $C_{k-1}$ observed from the body $C_0$.

**Remark 7.** *For* $m = 2$, ${}_0\underline{\boldsymbol{\omega}}_2 = \underline{\boldsymbol{\omega}}_1 + \underline{\boldsymbol{\omega}}_2$, *we will obtain the space replica of* ***Aronhold-Kennedy Theorem:*** *the instantaneous screw axis for the three relative rigid body motions has in every moment a common perpendicular, at any given time. The common perpendicular is line that corresponds to the dual vector* $\underline{\boldsymbol{\omega}}_1 \times \underline{\boldsymbol{\omega}}_2$.

To determine the field of the $n^{th}$ order accelerations of a rigid body $C_m$ we have to determine the ${}_0\underline{\boldsymbol{\omega}}_m^{(n)}$, $n \in \mathbb{N}$.

We denote $\underline{\boldsymbol{\omega}}_p^{[n]} = {}^0\underline{\mathbf{R}}_1 {}^1\underline{\mathbf{R}}_2 \ldots {}^{p-2}\underline{\mathbf{R}}_{p-1} \underline{\boldsymbol{\Omega}}_p^{(n)}$ the $\mathrm{n^{th}}$ order derivative of the relative dual twist $\underline{\boldsymbol{\Omega}}_p$, resolved in the body frame of $C_0$.

In order to determine the $\mathrm{n^{th}}$ order accelerations field of a rigid body $C_m$, we have to determine the ${}_0\underline{\boldsymbol{\omega}}_m^{(n)}$, $n \in \mathbb{N}^*$.

To compute ${}_0\underline{\boldsymbol{\omega}}_m^{(n)}$, $n \in \mathbb{N}^*$ we will use the following

**Lemma**: *If* $\underline{\boldsymbol{\omega}}_{\mathrm{p}} = \underline{\mathbf{R}}\, \underline{\boldsymbol{\Omega}}_{\mathrm{p}}$ *with* $\underline{\mathbf{R}} \in \underline{\mathbf{SO}}_3^{\mathbb{R}}$ *and* $\underline{\boldsymbol{\omega}}_{\mathrm{p}}, \underline{\boldsymbol{\Omega}}_{\mathrm{p}} \in \underline{\mathbf{V}}_3^{\mathbb{R}}$, *then*

$$\underline{\boldsymbol{\omega}}_{\mathrm{p}}^{(\mathrm{n})} = \mathbf{p}_{\mathrm{n}}(\underline{\boldsymbol{\omega}})\underline{\boldsymbol{\omega}}_{\mathrm{p}}, \mathrm{p} = \overline{1, \mathrm{n}} \tag{123}$$

*where* $\mathbf{p}_{\mathrm{n}}(\underline{\boldsymbol{\omega}})$ *are polynomials of the differential operator* $\mathbf{D} = \frac{\mathrm{d}}{\mathrm{dt}}$, *with coefficients in the non-commutative ring of Euclidian dual tensors.*

$$\mathbf{p}_{\mathrm{n}}(\underline{\boldsymbol{\omega}}) = \sum_{\mathrm{k}=0}^{\mathrm{n}} \mathrm{C}_{\mathrm{n}}^{\mathrm{k}} \underline{\boldsymbol{\Phi}}_{\mathrm{n-k}} \mathbf{D}^{[\mathrm{k}]}, \tag{124}$$

*where* $\mathrm{C}_n^k$ *is the binomial coefficient,* $\mathbf{D}^{[\mathrm{k}]}\boldsymbol{\omega}_{\mathrm{p}} = \boldsymbol{\omega}_{\mathrm{p}}^{[\mathrm{k}]}$ *and* $\underline{\boldsymbol{\Phi}}_{\mathrm{p}}$ *are dual tensors*

$$\underline{\boldsymbol{\Phi}}_{\mathrm{p}} = \underline{\mathbf{R}}^{(\mathrm{p})}\underline{\mathbf{R}}^{\mathrm{T}}, \underline{\mathbf{R}} \in \underline{\mathbf{SO}}_3^{\mathbb{R}}, \mathrm{p} = \overline{0, \mathrm{n}}, \tag{125}$$

*which follow the recurrence equation*:

$$\begin{cases} \underline{\boldsymbol{\Phi}}_{\mathrm{p+1}} = \dot{\underline{\boldsymbol{\Phi}}}_{\mathrm{p}} + \underline{\boldsymbol{\Phi}}_{\mathrm{p}}\tilde{\underline{\boldsymbol{\omega}}} \\ \underline{\boldsymbol{\Phi}}_0 = \underline{\mathbf{I}} \end{cases}, \mathrm{p} \in \mathbb{N}. \tag{126}$$

**Theorem 11.** *The following equation takes place*

$$\begin{aligned} {}_0\underline{\boldsymbol{\omega}}_m^{(n)} = \underline{\boldsymbol{\omega}}_1{}^{(n)} + \mathbf{p}_n(\underline{\boldsymbol{\omega}}_1)\underline{\boldsymbol{\omega}}_2 + \mathbf{p}_n(\underline{\boldsymbol{\omega}}_1 + \underline{\boldsymbol{\omega}}_2)\underline{\boldsymbol{\omega}}_3 + \ldots \\ + \mathbf{p}_n(\underline{\boldsymbol{\omega}}_1 + \underline{\boldsymbol{\omega}}_2 + \ldots + \underline{\boldsymbol{\omega}}_{m-1})\underline{\boldsymbol{\omega}}_m; \forall n \in \mathbb{N} \end{aligned} \tag{127}$$

*where* $\mathbf{p}_n(\boldsymbol{\omega})$ *are polynomials of the derivative operator* $\mathbf{D} = \frac{d}{dt}$, *with coefficients in the non-commutative ring of Euclidian dual tensors*

$$\mathbf{p}_n(\underline{\boldsymbol{\omega}}) = \sum_{k=0}^{n} \mathrm{C}_n^k \underline{\boldsymbol{\Phi}}_{n-k} \mathbf{D}^{[k]} \tag{128}$$

*where* $\mathrm{C}_n^k$ *is the binomial coefficient,* $\mathbf{D}^{[\mathrm{k}]}\underline{\boldsymbol{\omega}}_{\mathrm{p}} = \underline{\boldsymbol{\omega}}_{\mathrm{p}}^{[\mathrm{k}]}$ and $\underline{\boldsymbol{\Phi}}_p$ *are dual tensors*

$$\underline{\boldsymbol{\Phi}}_p = \underline{\mathbf{R}}^{(p)} \cdot \underline{\mathbf{R}}^T, \underline{\mathbf{R}} \in \underline{\mathbf{SO}}_3^{\mathbb{R}}, p = \overline{0, n} \tag{129}$$

*which follow the recurrence equation*:

$$\begin{cases} \underline{\boldsymbol{\Phi}}_{p+1} = \underline{\dot{\boldsymbol{\Phi}}}_p + \underline{\boldsymbol{\Phi}}_p \underline{\tilde{\boldsymbol{\omega}}} \\ \underline{\boldsymbol{\Phi}}_0 = \underline{\mathbf{I}} \end{cases}, p \in \mathbb{N}. \tag{130}$$

Other equivalent forms of Eq. (127) are the following recursive formulas (**Figures 3** and **4**):

$$\underline{\boldsymbol{\omega}}_m^{(n)} = \underline{\boldsymbol{\omega}}_1^{(n)} + \mathbf{p}_n\left({}_0\underline{\boldsymbol{\omega}}_1\right)\underline{\boldsymbol{\omega}}_2 + \mathbf{p}_n\left({}_0\underline{\boldsymbol{\omega}}_2\right)\underline{\boldsymbol{\omega}}_3 + \dots + \mathbf{p}_n\left({}_0\underline{\boldsymbol{\omega}}_{m-1}\right)\underline{\boldsymbol{\omega}}_m, \forall n \in \mathbb{N} \tag{131}$$

The previous equations are valid in the most general situation where there are no kinematic links between the rigid bodies $C_1, C_2, \dots, C_m$.

The following identity can be proved:

$$\underline{\boldsymbol{\Phi}}_k\left(\underline{\boldsymbol{\omega}}_1 + \underline{\boldsymbol{\omega}}_2 + \dots + \underline{\boldsymbol{\omega}}_{p-1}\right) = \sum_{k_1+k_2+\dots+k_{p-1}=k} \mathrm{C}_n^{k_1,\dots,k_{p-1}} \underline{\boldsymbol{\Phi}}_{k_1}(\underline{\boldsymbol{\omega}}_1)\underline{\boldsymbol{\Phi}}_{k_2}(\underline{\boldsymbol{\omega}}_2)\dots\underline{\boldsymbol{\Phi}}_{k_{p-1}}(\underline{\boldsymbol{\omega}}_{p-1}) \tag{132}$$

where $\mathrm{C}_n^{k_1,\dots,k_{p-1}} = \frac{n!}{k_1!\dots k_{p-1}!}$ is the multinomial coefficient.

From Eq. (131), on obtain the closed form non-recursive coordinate-free formula:

$$\begin{aligned} {}_0\underline{\boldsymbol{\omega}}_m^{(n)} &= \underline{\boldsymbol{\omega}}_1^{[n]} + \underline{\boldsymbol{\omega}}_2^{[n]} + \dots + \underline{\boldsymbol{\omega}}_m^{[n]} + \\ &+ \sum_{p=2}^{m}\sum_{k=1}^{n} \mathrm{C}_n^k \sum_{k_1+\dots+k_{p-1}=k} \mathrm{C}_n^{k_1,\dots,k_{p-1}} \underline{\boldsymbol{\Phi}}_{k_1}(\underline{\boldsymbol{\omega}}_1)\dots\underline{\boldsymbol{\Phi}}_{k_{p-1}}(\underline{\boldsymbol{\omega}}_{p-1})\underline{\boldsymbol{\omega}}_p^{[n-k]}, \end{aligned} \tag{133}$$

where

$$\begin{aligned} \underline{\boldsymbol{\Phi}}_0(\underline{\boldsymbol{\omega}}) &= \underline{\mathbf{I}} \\ \underline{\boldsymbol{\Phi}}_1(\underline{\boldsymbol{\omega}}) &= \underline{\tilde{\boldsymbol{\omega}}} \end{aligned} \tag{134}$$

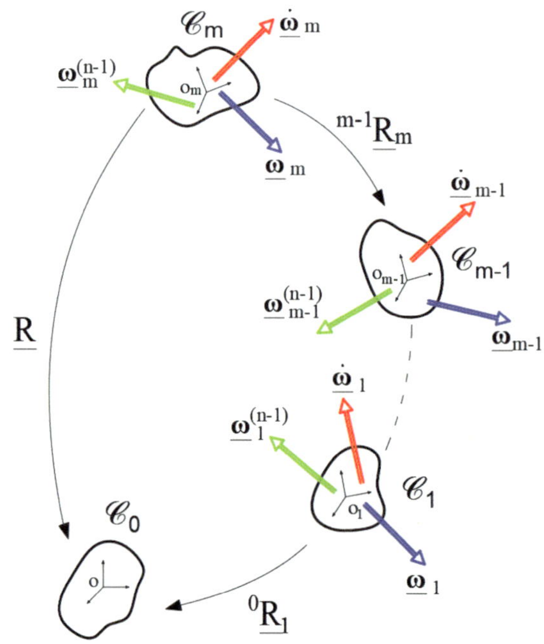


**Figure 3.**
*Higher-order time derivative of dual twist of relative motion.*

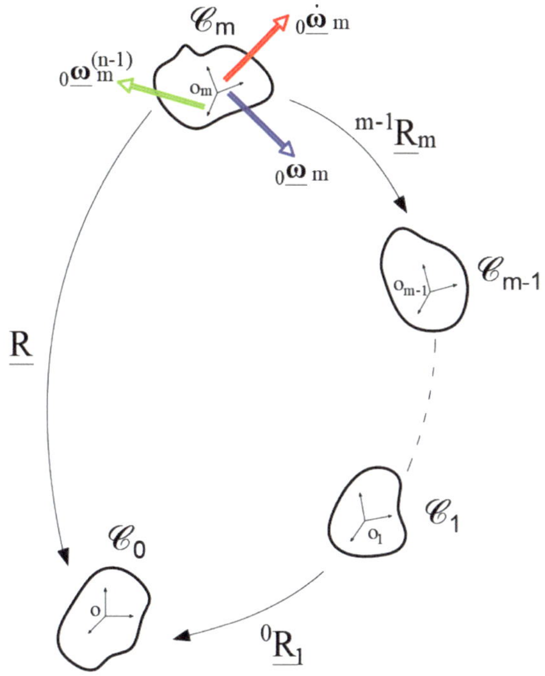


**Figure 4.**
*Higher-order time derivative of dual twist of relative motion on terminal body.*

$$\underline{\boldsymbol{\Phi}}_2(\underline{\boldsymbol{\omega}}) = \tilde{\underline{\boldsymbol{\omega}}}^{[1]} + \tilde{\underline{\boldsymbol{\omega}}}^2$$

$$\underline{\boldsymbol{\Phi}}_3(\underline{\boldsymbol{\omega}}) = \tilde{\underline{\boldsymbol{\omega}}}^{[2]} + \tilde{\underline{\boldsymbol{\omega}}}\tilde{\underline{\boldsymbol{\omega}}}^{[1]} + 2\tilde{\underline{\boldsymbol{\omega}}}^{[1]}\tilde{\underline{\boldsymbol{\omega}}} + \tilde{\underline{\boldsymbol{\omega}}}^3$$

$$\underline{\boldsymbol{\Phi}}_4(\underline{\boldsymbol{\omega}}) = \tilde{\underline{\boldsymbol{\omega}}}^{[3]} + \tilde{\underline{\boldsymbol{\omega}}}\,\tilde{\underline{\boldsymbol{\omega}}}^{[2]} + \left(3\tilde{\underline{\boldsymbol{\omega}}}^{[1]} + \tilde{\underline{\boldsymbol{\omega}}}^2\right)\tilde{\underline{\boldsymbol{\omega}}}^{[1]} + 3\tilde{\underline{\boldsymbol{\omega}}}^{[2]}\,\tilde{\underline{\boldsymbol{\omega}}} + 3\tilde{\underline{\boldsymbol{\omega}}}^{[1]}\,\tilde{\underline{\boldsymbol{\omega}}}^2 + 2\tilde{\underline{\boldsymbol{\omega}}}\,\tilde{\underline{\boldsymbol{\omega}}}^{[1]}\tilde{\underline{\boldsymbol{\omega}}} + \tilde{\underline{\boldsymbol{\omega}}}^4$$

$$\ldots$$

$$\underline{\boldsymbol{\Phi}}_n(\boldsymbol{\omega}) = \mathbf{P}_n\tilde{\underline{\boldsymbol{\omega}}}, \mathrm{n} \in \mathbb{N}^* \tag{135}$$

## 8. Higher-order kinematics for general 2C manipulator

We'll apply the general results obtained in the previous chapter for the particular case of four degrees of freedom 2C general manipulator. In this case the relative motions of three bodies $C_0, C_1, C_2$ are given, the spatial motion of the terminal body $C_2$ been described by dual orthogonal tensor as it follows:

$$^0\underline{\mathbf{R}}_2 = {}^0\underline{\mathbf{R}}_1{}^1\underline{\mathbf{R}}_2 \tag{136}$$

where

$$^0\underline{\mathbf{R}}_1 = e^{\underline{\alpha}_1(t)\,{}_0\tilde{\underline{\mathbf{u}}}_1} \tag{137}$$

$$^1\underline{\mathbf{R}}_2 = e^{\underline{\alpha}_2(t)\,{}_1\tilde{\underline{\mathbf{u}}}_2} \tag{138}$$

In Eqs. (138) and (139), the dual angles $\underline{\alpha}_1(t)$ and $\underline{\alpha}_2(t)$ are four times differentiable functions, and unit dual vectors $_0\underline{\mathbf{u}}_1$ and $_1\underline{\mathbf{u}}_2$ being constant. To simplify the writing, we will denote:

$$ {}_0\underline{\mathbf{u}}_1 = \cdot\underline{\mathbf{u}}_1 \tag{139}$$

$$ {}_0\underline{\mathbf{u}}_2 = \left(I + sin\,\underline{\alpha}_1{}_0\tilde{\underline{\mathbf{u}}}_1 + (1 -\ cos\,\underline{\alpha}_1)_0\tilde{\underline{\mathbf{u}}}_1^2\right)_1\underline{\mathbf{u}}_2 = \cdot\underline{\mathbf{u}}_2 \tag{140}$$

$$\underline{\omega}_1 = \dot{\alpha}_1(t) + \varepsilon\dot{d}_1(t) \tag{141}$$

$$\underline{\omega}_2 = \dot{\alpha}_2(t) + \varepsilon\dot{d}_2(t) \tag{142}$$

According to the observations from Section 6, the vector field of the velocity, the acceleration, the jerk, the jounce is uniquely determined by the dual vectors $\underline{\boldsymbol{\omega}}, \dot{\underline{\boldsymbol{\omega}}}, \ddot{\underline{\boldsymbol{\omega}}}, \dddot{\underline{\boldsymbol{\omega}}}$. Taking into account Eq. (133), we will have:

$$\underline{\boldsymbol{\omega}} = \underline{\omega}_1\underline{\mathbf{u}}_1 + \underline{\omega}_2\underline{\mathbf{u}}_2 \tag{143}$$

$$\dot{\underline{\boldsymbol{\omega}}} = \dot{\underline{\omega}}_1\underline{\mathbf{u}}_1 + \dot{\underline{\omega}}_2\underline{\mathbf{u}}_2 + \underline{\omega}_1\underline{\omega}_2\underline{\mathbf{u}}_1 \times \underline{\mathbf{u}}_2 \tag{144}$$

$$\ddot{\underline{\boldsymbol{\omega}}} = \ddot{\underline{\omega}}_1\underline{\mathbf{u}}_1 + \ddot{\underline{\omega}}_2\underline{\mathbf{u}}_2 + (2\underline{\omega}_1\dot{\underline{\omega}}_2 + \dot{\underline{\omega}}_1\underline{\omega}_2)\underline{\mathbf{u}}_1 \times \underline{\mathbf{u}}_2 + \underline{\omega}_1^2\underline{\omega}_2\underline{\mathbf{u}}_1 \times (\underline{\mathbf{u}}_1 \times \underline{\mathbf{u}}_2) \tag{145}$$

$$\ddot{\underline{\boldsymbol{\omega}}} = \ddot{\underline{\omega}}_1\underline{\mathbf{u}}_1 + \ddot{\underline{\omega}}_2\underline{\mathbf{u}}_2 + \left(\ddot{\underline{\omega}}_1\underline{\omega}_2 + 3\dot{\underline{\omega}}_1\dot{\underline{\omega}}_2 + 3\underline{\omega}_1\ddot{\underline{\omega}}_2 - \underline{\omega}_1^3\underline{\omega}_2\right)\underline{\mathbf{u}}_1 \times \underline{\mathbf{u}}_2 + 3\left(\underline{\omega}_1^2\dot{\underline{\omega}}_2 + \dot{\underline{\omega}}_1\underline{\omega}_1\underline{\omega}_2\right)\underline{\mathbf{u}}_1 \times (\underline{\mathbf{u}}_1 \times \underline{\mathbf{u}}_2) \tag{146}$$

$$\begin{aligned}\dddot{\underline{\boldsymbol{\omega}}} = &\left[\dddot{\underline{\omega}}_1 + 3\left(\underline{\omega}_1^2\dot{\underline{\omega}}_2 + \dot{\underline{\omega}}_1\underline{\omega}_1\underline{\omega}_2\right)\underline{\mathbf{u}}_1 \cdot \underline{\mathbf{u}}_2\right]\underline{\mathbf{u}}_1 + \left[\dddot{\underline{\omega}}_2 - 3\left(\underline{\omega}_1^2\dot{\underline{\omega}}_2 + \dot{\underline{\omega}}_1\underline{\omega}_1\underline{\omega}_2\right)\right]\underline{\mathbf{u}}_2 \\ &+ \left(\ddot{\underline{\omega}}_1\underline{\omega}_2 + 3\dot{\underline{\omega}}_1\dot{\underline{\omega}}_2 + 3\underline{\omega}_1\ddot{\underline{\omega}}_2 - \underline{\omega}_1^3\underline{\omega}_2\right)\underline{\mathbf{u}}_1 \times \underline{\mathbf{u}}_2\end{aligned} \tag{147}$$

Similarly, the results for six degrees of freedom general 3 C manipulator can be obtained, the calculus being a little longer.

## 9. Conclusions

The higher-order kinematics properties of rigid body in general motion had been deeply studied. Using the isomorphism between the Lie group of the rigid displacements $S\mathbb{E}_3$ and the Lie group of the orthogonal dual tensors $\underline{S\mathbb{O}}_3$, a general method for the study of the field of arbitrary higher-order accelerations is described. It is proved that all information regarding the properties of the distribution of high-order accelerations are contained in the n-th order derivatives of the dual twist of the rigid body. These derivatives belong to the Lie algebra associated to the Lie group $\underline{S\mathbb{O}}_3$.

For the case of the spatial relative kinematics, equations that allow the determination of the n-th order field accelerations are given, using a Brockett-like formulas specific to the dual algebra. In particular cases the properties for velocity, acceleration, jerk, hyper-jerk (jounce) fields are given.

The obtained results interest the theoretical kinematics, jerk and jounce analysis in the case of parallel manipulations, control theory and multibody kinematics.

## Author details

Daniel Condurache
Technical University of Iasi, Iasi, Romania

*Address all correspondence to: daniel.condurache@tuiasi.ro

## References

[1] Everett JD. On the kinematics of a rigid body. The Quarterly Journal of Pure and Applied Mathematics. 1875;**13**: 33-66

[2] Ball RS. Theory of Screws. London: Cambridge University Press Warehouse; 1900

[3] Hunt KH. Kinematic Geometry of Mechanisms. New York: Oxford University Press; 1978

[4] Phillips J. Freedom in Machinery: Introducing Screw Theory. Vol. 1. Cambridge: Cambridge University Press; 1984

[5] Karger A. Singularity analysis of serial robot-manipulators. ASME Journal of Mechanical Design. 1996; **118**(4):520-525

[6] Rico JM, Gallardo J, Duffy J. Screw theory and higher order kinematic analysis of open serial and closed chains. Mechanism and Machine Theory. 1999; **34**(4):559-586

[7] Lerbet J. Analytic geometry and singularities of mechanisms. Zeitschrift für Angewandte Mathematik und Mechanik. 1999;**78**(10):687-694

[8] Müller A. Higher derivatives of the kinematic mapping and some applications. Mechanism and Machine Theory. 2014;**76**:70-85

[9] Müller A. An overview of formulae for the higher-order kinematics of lower-pair chains with applications in robotics and mechanism theory. Mechanism and Machine Theory. 2019;**142**

[10] Condurache D, Matcovschi M. Computation of angular velocity and acceleration tensors by direct measurements. Acta Mechanica. 2002; **153**(3–4):147-167

[11] Angeles J. The application of dual algebra to kinematic analysis, computational methods. Mechanical Systems. 1998;**161**:3-31

[12] Bokelberg EH, Hunt KH, Ridley PR. Spatial motion-I: Points of inflection and the differential geometry of screws Raumliche bewegung-I wendepunkte und die differentiale schraubengeometrie. Mechanism and Machine Theory. 1992; **27**(1):1-15

[13] Condurache D, Burlacu A. Orthogonal dual tensor method for solving the AX=XB sensor calibration problem. Mechanism and Machine Theory. 2016;**104**:382-404

[14] Condurache D, Burlacu A. Dual tensors based solutions for rigid body motion parameterization. Mechanism and Machine Theory. 2014;**74**:390-412

[15] Condurache D, Burlacu A. Dual lie algebra representations of the rigid body motion. In: AIAA/AAS Astrodynamics Specialist Conference; San Diego, CA; 2014

[16] Condurache D. A Davenport dual angles approach for minimal parameterization of the rigid body displacement and motion. Mechanism and Machine Theory. 2019; **140**:104-122

[17] Condurache D. General rigid body motion parameterization using modified Cayley transform for dual tensors and dual quaternions. In: The 4th Joint International Conference on Multibody System Dynamics; Montreal; 2016

[18] Condurache D. Higher-order accelerations on rigid bodies motions. A tensors and dual lie algebra approach. Acta Technica Napocensis – Series: Applied Mathematics, Mechanics and Engineering. 2018;**61**(1):29-38

[19] Condurache D. Higher-order kinematics of rigid bodies. A tensors algebra approach. In: Kecskeméthy A, Flores FG, Carrera E, Elias DA, editors. Interdisciplinary Applications of Kinematics. Mechanisms and Machine Science. Vol. 71. Cham: Springer; 2019. pp. 215-225

[20] Condurache D. Higher-order relative kinematics of rigid body motions: A dual lie algebra approach. In: Lenarcic J, Parenti-Castelli V, editors. Advances in Robot Kinematics. Vol. 8. 2018. pp. 83-91

[21] Condurache D. Higher-order acceleration centers and kinematic invariants of rigid body. In: The 5th Joint International Conference on Multibody System Dynamics; Lisbon; 2018

[22] Condurache D. Higher-order Rodrigues dual vectors. Kinematic equations and tangent operator. In: The 5th Joint International Conference on Multibody System Dynamics; Lisbon; 2018

[23] Pennestrı E, Valentini PP. Linear dual algebra algorithms and their application to kinematics. In: Multibody Dynamics, Computational Methods and Applications. Dordrecht: Springer; 2009. pp. 207-229

[24] Samuel A, McAree P, Hunt K. Unifying screw geometry and matrix transformations. The International Journal of Robotics Research. 1991; **10**(5):454-472

[25] Veldkamp GR. Canonical systems and instantaneous invariants in spatial kinematics. Journal of Mechanisms. 1967;**2**(3):329-388